VALUE ENGINEERING

A Plan for *Invention*

VALUE ENGINEERING
A Plan for *Invention*

Richard J. Park, PE, CVS, FSAVE

R.J. Park and Associates Inc.
Birmingham, Michigan

S^t_L

St. Lucie Press
Boca Raton London New York Washington, D.C.

Library of Congress Cataloging-in-Publication Data

Park, Richard J.
    Value engineering: a plan for invention / by Richard J. Park.
        p. cm.
    Includes bibliographical references and index.
    ISBN 1-57444-235-X (alk. paper)
    1. Value analysis (Cost control)   I. Title.
TS165.P37   1998
658.15'52–dc21   98-36770
                                                                   CIP

No claim to original U.S. Government works
International Standard Book Number 1-57444-235-X
Library of Congress Card Number   98-36770
Printed in the United States of America   1 2 3 4 5 6 7 8 9 0
Printed on acid-free paper

The Author

Richard J. Park is president of R. J. Park and Associates, Inc., Birmingham, Michigan, and a member of the TECC Group. He has had extensive experience in Value Engineering (VE) since 1961, and in that time he has organized VE operations in three major companies in the automotive, aircraft, and general manufacturing fields and has conducted consulting operations since 1980. He has trained hundreds of people in industry workshops, college courses, and society activities and conducted studies that have resulted in substantial benefit for the client.

Mr. Park holds a Bachelor of Mechanical Engineering degree from Clarkson University and a Master of Mechanical Engineering degree from New York University. He is a registered Professional Engineer in the states of New Jersey, New York, and Michigan, and is a Certified Value Specialist. He is a fellow of the Society of American Value Engineers, which has recently changed its name to SAVE International, and a member of the Society of Manufacturing Engineers and the Engineering Society of Detroit.

In the course of his work he has developed and refined methods of operation that have expanded the potential of VE to a full management discipline. He has been recognized as a primary force in the development of the Function Analysis System Technique (FAST), of which he calls his version the ARGUS System, and the application of the system to a wide range of business, management, and technical problems that have resulted in innovation, cost reduction, and productivity improvement for his clients.

A major part of his present business is conducting programs to aid companies to improve productivity, product quality, performance, and cost. Much of this work has been conducted in joint programs staffed by members of both the customer and supplier companies to achieve and maintain profitable operations.

Prior to 1961, Mr. Park had extensive experience in the chemical, petroleum, and manufacturing industries where he managed and supervised activities involving new product line introduction and consulting services to provide management guidance on operational problems and major capital investments.

He has held numerous chapter and national offices in the Society of American Value Engineers (SAVE) including National Vice President of Professional Development. In recognition of his outstanding work for the Society and the profession, Mr. Park was named "Value Engineer of the Year" (1972–73). He is a recognized speaker on VE and has a number of publications to his credit. He received the award for the Best Technical Paper of 1979 and was elected a Fellow of the Society of American Value Engineers in 1980. He has been an instructor in VE at several universities and an assistant professor at Michigan State University.

Preface

When Albert Einstein said that he was reminded a hundred times a day how his life was dependent upon the contributions of other men, he was pointing out a major fact of life.

As I developed this text, the contributions of those whom I have met and worked with along the way were brought to mind and I realized how much they contributed to my knowledge and skills and how these contributions helped me to do a better job. In most cases, these contributions were positive; then at other times, they pointed out shortcomings that caused me to look deeper or in different directions.

This awareness was highlighted the more deeply and more involved I became in Value Engineering (VE). I began to realize the broad range of contributions I have received from other people and how much they helped to broaden my viewpoint to effect a successful VE operation. In developing and conducting our operations at the Chrysler Corporation, there was Ed Dobzyniak, our cost specialist; George Lumsden, our pioneer in listening techniques; Joe Picraux, whose manpower development lectures brought up subjects that most engineers rarely consider; Dick Darios, who combined creativity and psychology in an educational and entertaining way; Tom Smith, our motivational specialist, who always created a desire to learn more; and Bob Lehman from the Chrysler Management Institute, who guided us to a successful organization. In addition, we had the support and understanding of Tex Hearon and Phil Kornblum, our controllers, whose interest and enthusiasm helped us to develop our methods from product-oriented applications to technical and administrative operations and often into fuzzy areas that were not clearly defined.

After I left Chrysler and entered the consulting field, Paul Frusti, who worked with me to develop our Chrysler operation, continued to provide his technical and manufacturing and creative know-how as my associate and made major contributions in the field.

When I decided to put this book together I determined that it should cover, at least in part, the broad range of knowledge required for successful VE application. VE is not a cookbook procedure. It requires thinking in a wide range of knowledge. However, writing the text was one thing; packaging the information into a practical unit was another. Ed Heussner became my computer consultant, and his advice was invaluable in dealing with the electronic mind. The project also became a family affair, with my son, Rick, aiding in producing many figures and diagrams, and my daughter, Chris, contributing subjective input, but most of all my wife, Barbara, who supported me both physically and psychologically to complete the project.

I thank you all for your contributions and hope that the information in this book will help others in some way as you have helped me during the lifetime of collecting this useful store of knowledge.

Contents

SECTION III: THE HUMAN ELEMENT

APPENDIX

THE WORLD
AROUND US

Chapter 1

Background: From Fire to Flight

Early man did not have a great background of information to draw upon, but he survived and advanced. He was alert to his surroundings and his needs and somehow was able to put them together. Thus he observed fire and recognized its benefits. In time he learned to make clothes and invented the wheel. Bridges probably developed from fallen logs; then came roads to make the wheel work better. Improvements on the wheel eventually evolved into machines. All of this development came with very little information to draw upon. Even today "You have to look hard to find cases in which the theory is well worked out before the practice."[1] For example, the steam engine came before thermodynamics and caissons were used for bridge footings before there was an understanding of the cause of the "bends."

Yet man has been successful, not only in surviving but in advancing society to the highly developed state in which we live today. Today our society is constantly being bombarded by information from all sides. Radio, television, the internet, and magazines of all kinds are available for entertainment as well as education. With all of this information to draw upon we should be able to solve all of our problems with ease. However, such is not the case. As our information becomes greater, our problems become more complex.

Every day millions of dollars and hundreds of thousands of man-hours are being wasted by highly educated people working diligently to solve the wrong problems. This state of affairs is brought about for many reasons; among them is the shortage of time, manpower, and money. The result is that the problem must be solved immediately; action is required. There is no time to think when action is required. This not only adds extra cost to our products and services but wastes time that could be put to better use for the benefit of society. It is my strong belief that there are many important things that are not being done because we are so busy doing things that do not have to be done. If we could identify these unnecessary activities, we could then apply the wasted effort to the things that are not being done for an overall benefit.

Of course, every effort must be made to correct or eliminate problems as soon as possible so that operations can return to normal; however, experience has shown that the problem may not be what it seems and immediate action without proper analysis may only eliminate the symptom of the problem, and the problem may still remain. This creates additional problems in that it always increases cost and frequently affects product quality, reliability, and overall performance.

This situation exists because it really is not easy to identify a problem at first glance. One of the primary reasons is that, unlike early man who was totally alert to his surroundings, we have become specialists in a specific field. This makes it difficult for us to see beyond our special area of expertise. In fact, in many cases it causes us to try to make the problem one we believe we can solve rather than analyzing the situation to determine just what is required to return to normal. In other words, people see what they expect to see and ignore what they don't expect. This condition is a phenomenon that psychologists call perceptual bias.

At one time in my career I was working for a foreign oil company whose operations were in the Middle East. It was a time when the industry was moving to the use of larger and larger supertankers, and it was necessary to build new loading facilities to accommodate them. The existing facilities consisted of a long pier from the shore about a half mile into the gulf to deep water, with a large extension on the end for tanker docking. The entire structure was in the shape of a T. All of the loading facilities were on the T part of the structure. However, over the years we found that the particular design was impossible to maintain economically because of the highly corrosive property of the very warm salt water and the significant tidal condition.

I was asked to review the design and make a proposal for a facility that would meet the new requirements and be more efficient from the standpoint of performance, maintenance, and cost.

I reviewed the design thoroughly from the standpoint of my background and knowledge, developed a new, more efficient design concept, found new, more effective ways to satisfy performance requirements, and found that the cost was only half that of a facility built on the old concept.

My recommendations were well received at the local level and supported by the most prestigious port facility design consultant in the country. However, the design was rejected by the field because it was different, so the new facility was built to the same old standards and carried over all of the old problems.

This was a very big disappointment to me. Why wouldn't management accept this new lower cost facility that obviously appeared to be better? Tradition, ego, trust, whatever the reason, it was a distressing question.

Several years later, when I was working for a different company, I found that a major construction periodical's lead article was on new offshore loading facilities at my former company. I immediately turned to the article to see if they had adopted my advanced design and found that they had jumped completely over me to an entirely new concept. The new design consisted of a causeway dredged from the sea and a series of small artificial islands connected by walkways. All of the necessary loading equipment was installed on the islands. There was no pier at all. A very simple solution. Why didn't I think of that?

In this case the answer came to me immediately. I was asked to come up with a better pier design, which I did. No one asked me to reformulate the problem. I did not ask myself what the problem was; I simply went down the old path — perceptual bias. This caused me to reconsider the situation, and I found some interesting things about standard practices. For example, most designs are based on what was done the last time. Whether it's piers, oil refineries, chemical plants, automobile parts and assemblies, or organizations, we look only to see how the job was done and are guided by the past.

It is my belief that the solution to more effective thinking lies within every one of us. There is no system that will help us to think more effectively if we do not have the desire to do a better job. However, if we are interested in improving our thinking processes, there are a number of systems that can help us.

This is a book about Value Engineering (VE). I have been in the VE business for over 40 years and have come to believe that most people look at VE as a follow-the-steps procedure to cost reduction. However, over the years I have conducted college courses and industry workshops in the system and believe VE is a process to open the mind to new ideas by breaking down constraints to visualization. We all have a tendency to go by past experience. If it worked in the past, it will work in the future. This is true; however, following this path leads us down the same road we have always trod. How do we break away from the old tried-and-true habit to develop new exciting methods, products, and inventions to carry us into the future ahead of the competition?

VE is a method that helps us to do exactly that. It helps us to break down our mental constraints and to open our minds to new and different methods that enable us to see our requirement in a different light. The means to do this was invented by Lawrence D. Miles[2] and is based on what he called function, or what the product or requirement *does* rather than what they *are*. As Larry said many times, "Ah functions! We buy only functions, what things do. It is the difference between a word and an idea." The development of VE through application over the years has made it possible for us to extend the process from products to services of all kinds and to any want or need, whether it's a system related to community affairs or a personal requirement.

Sometimes the term *function* can be misleading to people. However, some have construed the term *function* to mean purpose. They look upon function as the purpose of a product, service plan, or action, whether it is business or personal, a fact that many people seem to have overlooked. Function can help us to be better engineers, business people, or just help us to do a better job whether for an employer or ourselves.

However, if we are to expand the application of VE beyond products and services we must begin to broaden our total horizon as well. We must move beyond products, cost and value into the broader areas of human relations and creativity. These are areas in which many of us have little experience or education. For this reason I have divided this text into four sections, each covering the major aspects of the technical and social requirements and the application and examples of results in a broad area of application.

The first section, The World Around Us, refreshes our minds to systems of the past and present and reminds us of business systems and information that has been available to us for years. Some of the systems are no longer very popular; in fact some are no longer in use at all, and some are de rigueur until the next new system comes along. However, all of the systems have some beneficial attributes that can aid in organizing, defining, identifying, and improving overall operations, whether in manufacturing plants, the engineering process and even in your church, school group, or community affairs. This section also evaluates each system, identifies the strengths of the various systems, and shows where they can be applied to provide the most benefit in the project development cycle.

The Toolbox, in Chapter 3 is a matrix that identifies each management method and the level of benefit it can provide relative to 20 attributes. We have also identified the most beneficial time to apply each of the various methods in an Argus Chart, a method developed to aid in analyzing projects by applying the Theory of Function Relationships, a logic system, rather than by a sequential flow chart. The construction and use of the Argus Chart are explained in detail in Chapter 12, Value Engineering — A Total System.

The second section, The Economics of Profit, covers technical aspects of project development, whether a product or service, or a process or procedure for a business or for a voluntary school or community operation, or the ultimate need, for our family or self. The development of anything involves cost, whether it is in dollars and cents or time and effort. We spend money to develop a product and carefully measure the expenditure; however, we also spend time and effort on volunteer projects which have a cost in time and effort. Cost, whether in dollars or time and effort is still a cost, and we don't like to waste time any more than money, so we should strive to do the most efficient job possible.

The third section, The Human Element, is devoted to subjects sometimes called the soft sciences. Communications, motivation teams, and teamwork and creativity are necessary elements of the VE process that must be considered because they are important when working with people whether in teams or on an individual basis. Many have wondered why some people move up through an organization faster than others, even though they appear to have the same qualifications. Perhaps it's their human relations attributes. Part of this may be that they have learned to be better salesmen. An engineer rarely looks upon himself as a salesman but he should. He is always trying to sell ideas. However, I have heard it said that success is not for the one with the great idea but for the one who sells it to the world. Creativity may help us to get the great idea but a great idea is nothing more than a great idea unless it becomes something someone wants or needs.

Although creativity is a part of this section, it is really a different attribute. Creativity is the means to advance civilization into the future. However, in most cases we have become so wrapped up with the present we forget about the future. As a result we have tended to stifle our creativity with our knowledge and security. This section discusses the elements of creativity and offers some insights as to how to recover our creative abilities.

Section four, The Sum of the Parts, is the objective of the book. In this section we bring all of the elements of history, technical knowledge, and human relations together in VE. This is the only system that not only brings all of these elements together into a workable system but requires their use to achieve the best result from our resources, whether they be dollars or effort. The section discusses elements exclusively developed to improve the effectiveness of VE. It demonstrates the use of the systems by their application to a simple successful product project. This section also discusses the basic requirements for an organization to successfully apply VE to achieve maximum benefit. The final part of this section is a series of successful examples ranging from products and services to personal needs.

Although the application of VE is growing around the world and is presently being applied vigorously in Asia, from Japan and Korea to China and India, in every country in Europe, from the European Economic Community to Russia, Eastern Europe, and the Mid East, and in Africa, Australia, and New Zealand, we believe that in the U.S. it is an overlooked resource. We believe that every manager, engineer, and technical specialist can improve his effectiveness by learning how to apply function analysis, The ARGUS system and their creative abilities whether employed in a specific VE operation or if only to apply the lessons discussed here for one's own benefit.

References

1. Vincenti, W., What engineers know, *Invent. Technol.,* Winter, 20, 23, 1997.
2. Miles, L. D., *Techniques of Value Analysis and Engineering,* McGraw Hill, New York, 1961.

Chapter 2

Management Systems: Tools for the Toolbox

Is it better to strive to make the wrong part the most efficient way it can be made or is it better to make the right part the best way possible? This is the conflict between several of today's pet management systems.

Over the years dozens of management systems have been developed to help define problems and to create better organizations that operate more efficiently, improve the operation and result in increased profitability. Among the more popular and those that are still found to some degree are Management By Objectives (MBO), Zero Based Budgets (ZBB), Kepner-Trego (KT), Taguchi Methods (TM), Quality Function Deployment (QFD), Design for Manufacturing and Assembly (DFMA), Failure Mode Effect Analysis (FMEA), Kaizen Methods (KM), Total Quality Management (TQM) and its various derivatives, Value Analysis (VA) also called Value Engineering (VE), Value Management (VM) and several other similar titles, and the Theory of Inventive Problem Solving (TRIZ).

This list is not intended to be complete but only to offer some insights into some of the more popular systems and discuss some of them briefly to show their strong points and perceived weaknesses. The intention is also to show that none of them is a complete system to solve all problems or to operate a company, large or small. Each contains elements that, when properly understood and properly applied, can improve the operation of any organization.

Although the primary objective of this book is to create an understanding of the principles and application of VE, it is believed that it is important to understand that no system stands completely alone and the expectation that it does is a primary reason for the decline in interest in most systems. There is no pill that can be taken to improve operations, products, or services. It requires an understanding of how things are working, where problems are, and where to apply effort to achieve maximum benefit. One tool will not fit all cases. For this reason we must understand the tools that are available to us and learn to apply them so that we can apply the most effective tool to the situation. Unfortunately, it seems that once we develop a skill we are reluctant to give it up to try a new method. However, for best performance on the job we must carry a bag of tricks just as the mechanic carries a box of tools.

In this chapter we have drawn a picture of current conditions and have identified management tools that have seen considerable popularity over the past several decades. There are many more that have come to the forefront. However, the conditions identified in the text can be expanded to any of them once we see that there are differences that make each one more applicable to a particular situation than another. Some may fit several conditions, but none of them has equal strengths in all conditions.

Management Methods

Management by Objectives (MBO)

During the last half of the 19th century many managers learned that it was a major feat to organize a large organization and it was an equally difficult task to manage it properly for survival. They learned that no single individual could direct the activities of a large corporation, but if he could control the results he could manage even the largest.

These ideas were applied to appraisal systems where stated goals rather than personality traits defined appraisal criteria in a number of large corporations. However, it was seen to be a personal development gimmick. Sometime later George Odione[1] developed the simple appraisal system into a corporate systems approach and MBO was born.

MBO was designed to emphasize objectives and to provide steps to improve overall operations. The MBO system clarifies the following elements:

1. *Input* — Resources committed to an idea to produce a result. This includes capital, labor, and material.
2. *Activities* — Behavior of people. This includes designing, making, selling, accounting.

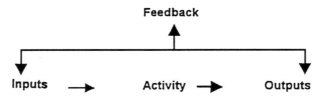

Figure 2.1

3. *Outputs* — goods and services that are the result of the combination of inputs and activities. The outputs are more valuable than all of the inputs, as shown in Figure 2.1.

MBO is a system where the manager and subordinate sit down together at the beginning of a period and talk until an agreement is reached on specific job goals. At the end of the period results are jointly reviewed and compared to goals. An assessment of results is made, adjustments agreed upon, and the process begins again.

It is claimed that MBO should accomplish the following four results:

1. Keep the goal in focus without becoming enmeshed in activities that mask the goal.
2. Clarify the role conflict between manager and subordinate at all levels.
3. MBO should be associated with the overall success of the organization.
4. Individual performance should improve because the individual has helped to define objectives and understands his goal.

The overall objective of MBO is excellent and still required. In addition, the steps to implement the process are clear and readily understandable. However, there are several major problems associated in successfully implementing a successful MBO operation.

1. Defining objectives is not an easy task.
2. Many people confuse output with input and become completely entangled in the activity trap.
3. Sitting down with subordinates and discussing operations until a goal is defined takes time, although most will agree it is the right thing to do.
4. The entire process requires change in the way things have been conducted in the past.

Peter Drucker[2] says, "MBO works if you first think through your objectives. Ninety percent of the time you haven't."

In all it breaks down to, "It's a good process but we just don't have the time." So, we go back to the same way things have been done in the past.

Zero-Based Budgets (ZBB) [3,4]

I can't think of anything more important in business than a zero-based budget (ZBB). One of the major problems in the automobile industry has been the use of incremental costs in product design. For example, when establishing the budget for a new model product an incremental cost over the old model may be set as a new overall target cost. The overall incremental cost is then broken down as a goal for each system and component. For example, a +$2.50 incremental cost over the old model may be set for the new model. However, the designer doesn't know if the old model system cost $2.50, $25, $250, or some other number. As a result, over a period of time he loses sight of what his product actually costs. He may not know if a bolt costs two cents or two dollars, and the entire system gets out of line.

The very same thing can happen to a corporation when incremental costs are used, as happens to a product. Costs continue to rise as all of the old problems are carried over into the future. Originally a ZBB was designed to be applied to overall operations. A ZBB says this is how much it will cost to run this operation in the coming period. It also establishes alternatives that can be substituted or added to the base budget to add new products or services. However, they must be justified with an incremental budget that says what you will get for the additional cost. These incremental costs are known as decision packages.

Just as with MBO, ZBB takes time to work out the goals and objectives of both the base cost and the cost of decision packages. It also requires that a five-step process be followed as shown below:

1. Identify objectives.
2. Evaluate alternative means to accomplish each objective.
3. Evaluate alternative funding levels.
4. Evaluate workloads.
5. Establish priorities.

Implementing this procedure requires that each alternative be defined as a decision unit, that the decision units be ranked and weighted, and that detailed operating budgets be prepared for each decision unit. Then the base must be defined and all units must be evaluated and built into a budget. This base unit and decision increments make it possible to build a total budget to accomplish the overall goal.

In discussing ZBB with company representatives I learned that their experience proved that the system was too complex to apply to the entire company but that it did work very well for R&D operations and individual R&D projects. I believe that many of the same problems associated with MBO are present in establishing a ZBB operation in a company. It is a new way of operating, involving much discussion and adjustment of requirements which appears to require too much time. So, we go back to the same old way.

The Kepner–Trego (K–T) System [5,6]

The K–T System is basically a decision-making process. The process provides a uniform method for solving problems, making decisions for action and methods to analyze a situation to determine potential problems. The methods are taught in a five-day workshop applied to a case problem. The process is broken down as follows:

1. Problem Analysis (PA)
2. Decision Analysis (DA)
3. Potential Problem Analysis (PPA)

The process recognizes that most managers do not make systematic use of information available to them to precisely identify a problem. In addition, they rarely make tests to determine possible causes for a problem, define the goal to be accomplished from any action taken, or to forecast the possible outcome. The K–T process leads a team of people through a series of steps and questions to clearly define a problem and achieve successful results from actions.

K–T defines a problem as a discrepancy between what is expected and what is taking place. If there is no discrepancy, there is no problem. This definition is expressed in Figure 2.2.

The system is implemented by following the series of questions and answers below:

1. Identify the Problem
      Recognize the problem
      Specify the problem and obtain information
      Find the cause

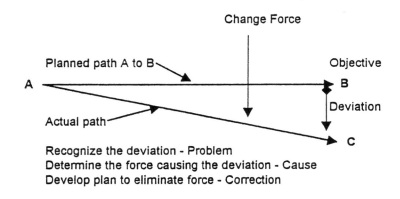

Change Force

Planned path A to B

Objective

A   B

Deviation

Actual path

C

Recognize the deviation - Problem
Determine the force causing the deviation - Cause
Develop plan to eliminate force - Correction

Figure 2.2

2.   Determine Course of Action
        Set objectives
        Consider alternative courses of action
        Choose the best course
        Implement and control the decision by getting feedback

A major step in the process is collecting information. Every effort must be made to find out as much as possible regarding the apparent problem. It is also necessary to be assured that the information is accurate and timely to avoid going down the wrong path By following a systematic series of specific questions, the team is guided by the system to a final conclusion.

Over 500 companies have conducted K–T workshops by certified K–T instructors. Many participants are enthusiastic about the results achieved and state how the process increased their awareness and opened their mind to alternative solutions. The K–T system is aimed directly toward the solution of specific problems. It does not affect overall operations as is the case with both MBO and ZBB. However, like MBO and ZBB, implementing the process does require a major change in corporate culture. It is team oriented and again takes time to collect information and conduct the analysis. Again, like both MBO and ZBB, it focuses on goals.

Taguchi Methods (TM)

One of the more successful new management systems is Taguchi Methods,[7] developed by Dr. Genichi Taguchi while he was working as an employee of the Nippon Telephone company. His major contribution is the combining of engineering and statistical methods to achieve rapid improvements in cost and quality by optimizing product design and manufacturing processes. He has developed the "Quality Loss Function," which makes it possible to calculate the cost of tolerance deviation, and the "Signal to Noise Ratio," which provides a quality measure early in design.

Taguchi Methods were introduced into the United States in 1982 with a number of large companies. TM is a specific application of Statistical Process Control (SPC) that emphasizes quality improvement through the identification and constant reduction of tolerances. Taguchi's theories state that designs should be "robust" enough so as not to be affected by manufacturing variables. Application requires constant monitoring of the process to identify deviations as soon as they occur, methods to eliminate the variables, reduce the tolerance, and repeat the cycle if necessary. Theoretically, the objective is to constantly reduce the variables and tolerances in the system to zero without adversely affecting product performance. Many statistical experts do not agree with his methods; few understand the theory, but most agree it works.

TM is a refined system for identifying and reducing manufacturing variables to improve quality and reduce cost. It is true, as Dr. Taguchi says, that all quality problems can be traced back to the design. It is also true that quality problems increase cost, and as a result, all cost problems can be traced back to the design. The system makes use of charts and equations to identify problem areas, uses the deviations to determine their cost, and employs design of experiment techniques to improve the product or service. Among the techniques claimed to be new are brainstorming, up-front thinking, and cooperative team effort.

TM is a complete system. Many of the methods have been assigned provocative names that tend to indicate that they are completely new. Tried-and-true statistical methods and systems are introduced, polished, and refined for a specific purpose and, as a total system, identify problems and aid in their solution. For example, Loss Function is a qualified deviation from a target. It is predictable and can be calculated using the system. It says, "Here's what it costs to vary from the target." The Loss Function creates a strong motivation to eliminate variables in a design.

The simple diagrams in Figures 2.3, 2.4, and 2.5 illustrate the TM objectives and illustrate the Loss Function theory.

To improve on a design, Taguchi recommends the Parameter Design System (PDS). Rather than solving the problem one step at a time, the PDS makes use of a number of statistical tools to find combinations of factors that will produce a design that is least affected by manufacturing variables. The system also provides for the collection and analysis of data

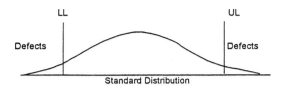

Figure 2.3   Normal Operations.   Reprinted from Byrne, D., and Taguchi, S., *The Taguchi Approach To Parameter Design,* **American Supplier Institute, Livonia, MI, 1986. With permission.**

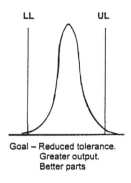

Figure 2.4   Improved Operations.   Reprinted from Byrne, D., and Taguchi, S., *The Taguchi Approach To Parameter Design,* **American Supplier Institute, Livonia, MI, 1986. With permission.**

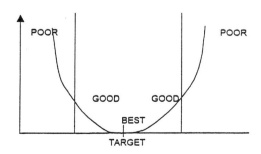

Figure 2.5   Loss Function.   Cost increases quadratically as deviation increases. The value of tolerance can be determined as goal for achievement. Reprinted from Byrne, D., and Taguchi, S., *The Taguchi Approach To Parameter Design,* **American Supplier Institute, Livonia, MI, 1986. With permission.**

and methods to aid in eliminating variables in the manufacturing process and for the creation of designs that are least affected by manufacturing variables. However, it does not offer methods for creative opportunities.

Quality Function Deployment (QFD) [8]

Like TM, Quality Function Deployment (QFD) is a statistical method aimed at improving product and process design before manufacture. QFD was first employed by Mitsubishi's Kobe shipyard in 1972. It introduces a disciplined, highly structured method to translate the customer's requirements into detailed technical language and targets. QFD helps rank critical items to determine when to apply major effort and TM. In effect TM is often a part of the QFD system. QFD is being used, at least in part, in the U.S. and Japan in both manufacturing and retail industries.

As powerful and as effective as TM is, in many cases it is being used in conjunction with QFD to determine where to apply TM. QFD brings the customer's requirements to bear throughout the entire organization, especially to the designers. It is a very disciplined system that makes use of a sophisticated set of charts to identify where to concentrate engineering time and effort.

It is claimed that QFD:

Reduces the number of engineering changes.
Shortens design cycles.
Lowers start-up costs.
Reduces warranty claims.
Improves customer satisfaction.
Creates competitive advantages.
Fosters Simultaneous Engineering (SE)

QFD makes use of a matrix to compare customer requirements to engineering solutions. It identifies the conflict and compatibility of engineering solutions to satisfy customer requirements. It can also be used to develop competitive analytical standings and a number of other beneficial data. A simplified example of the QFD planning matrix is shown in Figure 2.6.

It is claimed that results from QFD have been great. It is an excellent data collection system. It translates customer wants and needs to design, compares solutions to the customer requirements, and indicates conflicts. However, it does not do anything to aid in developing alternative creative designs.

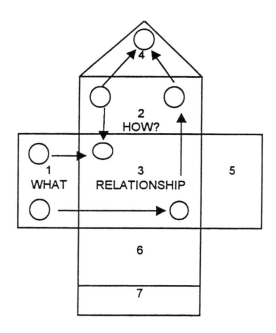

Figure 2.6   QFD Planning Matrix.   (1) Consumer requirements What? (2) Methods to satisfy requirements — How? (3) Relationship 1 to 2. (4) Compatibility and/or conflict of solutions 2. (5) Consumer rating of wants related to competition. (6) Performance values. (7) Competitive assessment. Reprinted from Sheller Globe Corporation, *Continuous Quality Improvement Handbook*, American Supplier Institute, Livonia, MI, 1983. With permission.

Kaizen

Kaizen emphasizes unending continuous improvement. This is another Japanese program that started in 1950 and is currently being applied vigorously by many American companies. One of the three major automobile companies has required their suppliers to attend Kaizen training programs, and they then monitor the companies to assure application. Kaizen looks to the workers to come up with ideas to improve their operations. Thus, it fosters teamwork. There is no particular procedure to follow; however, a facilitator leads the group through the process. The "team" usually has the authority to implement changes almost immediately.

Reports indicate that it has been very successful in most cases. However, there is no attempt to affect the product, only the process. The greatest complaint about the method is that it works to perfect the operation but does nothing about the design. In effect it may result in the very best way to produce a bad design.

Design for Assembly (DFA)[9]

Design for Assembly (DFA) or Design for Assembly and Manufacturing (DFMA) is again a Japanese system brought to America. DFA follows a prescribed procedure and a set of specific questions and equations designed to provide numerical rating for a design. The DFA system is and, in some cases, combines time and motion study and Value Engineering. The first step is to obtain concept sketches and a prototype sample. Determine the assembly method and assemble the product to determine assembly and insertion times using the prescribed charts. One chart determines the material handling times, and a second chart determines insertion times. These times are then posted on a third chart that provides a summary of the information developed. By using the information in the table, a simple calculation makes it possible to determine the theoretical number of parts in the design and the design efficiency. If the design efficiency is 70% or less, review the forms and determine the step in the process that requires excessive time, and then apply VE and creativity to reduce all of the major time elements.

It is expected that DFA will help:

Reduce the number of parts.
Reduce the number of required drawings.
Shorten development time.
Lower assembly cost.
Reduce engineering changes.
Reduce warranty.

DFA can be a major tool in product design, not only to simplify the design but also to simplify the manufacturing process. However, DFA does not provide a way to develop a better product. It follows a prescribed formula based on an existing design.

Total Quality Control (TQC)

A recent cartoon in an influential business publication shows a group of executives sitting around a table. A wall chart behind them shows profit heading straight down and a gleaming new toaster on the table. The caption reads, "We put the quality in before we found out how much it would cost."

This is no joke; it is a fact. Many companies have found out that quality can cost you money if you don't watch what you are doing.

A total quality program may be known by many names such as Total Quality Control (TQC), Statistical Process Control (SPC), Continuous Process Improvement, or others. The quality movement in the U.S. again comes to us from Japan. However, it is based on the theories of Dr. W. Edwards Deming,[10] an American who formulated his theories in the U.S. but was not able to sell his ideas to American management. Following World War II, the Japanese were looking for ways to put their economy back together and toured the world for ideas. In their search they discovered Dr. Deming and asked him to come to Japan to teach them his theories and methods. This he did so well that the Japanese have become world leaders in quality manufacturing. In fact, one of the highest awards a person can receive for contributions to manufacturing in Japan is the Deming Award.

At the 1981 SAVE International Conference in St. Louis, Mr. Shoichi Akazawa,[11] Executive Vice President of Fujitsu, said, "We didn't do anything particularly remarkable. We merely introduced your ideas and implemented them in our own way. Why don't you go home and practice what you preach." In this case we did. There are three quality philosophies in the U.S., Deming's theories, those of Dr. Joseph M. Juran,[12] and those of Philip B. Crosby.[12]

There is no question that quality theories, when properly implemented can have a positive effect in improving products. However, as pointed out above, quality is not always free, although it should be. One of the problems is losing sight of the fact that quality methods are not the goal. The goal is a product recognized as meeting the customer's expectations.

A TQC program is considered to be an overall company operation. Although the top quality authorities may not apply the same methods, they all do agree that in order to achieve success top management must be permanently involved. To create an effective Total Quality system in a company, Dr. Deming lists the following 14 obligations of top management:

Fourteen Obligations of Top Management

Dr. W. Edwards Deming

> MANAGEMENT PRINCIPLES
> 1. Innovate and allocate resources to fulfill the long-range needs of the company and customer rather than short-term profitability.
> 2. Discard the old philosophy of accepting defective products.

8. Reduce fear throughout the organization by encouraging open, nonpunitive communication. The economic loss resulting from fear to ask questions or report trouble is appalling.
9. Help reduce waste by encouraging design, research, and sales people to learn more about the problem of production.
14. Make maximum use of statistical knowledge and talent in your company.

TRAINING

6. Institute more thorough, better job-related training.
7. Provide supervision with knowledge of statistical methods; encourage use of these methods to identify which defects should be investigated for solution.
12. Institute rudimentary statistical training on a broad scale.
13. Institute a vigorous program for retraining people in new skills to keep up with changes in materials, methods, product designs, and machinery.

IMPLEMENTATION CONCEPTS

3. Eliminate dependence on mass inspection for quality control; instead, depend on process control through statistical techniques.
5. Use statistical techniques to identify the two sources of waste — system (85%) and local faults (15%); strive to constantly reduce this waste.
10. Eliminate the use of goals and slogans to encourage productivity, unless training and management support are also provided.
11. Closely examine the impact of work standards. Do they consider quality to help everyone to do a better job? They often act as an impediment to productivity improvement.

SUPPLIERS

4. Reduce the number of multiple-source suppliers. Price has no meaning without an integral consideration for quality. Encourage suppliers to use statistical process control.

Some of the benefits that can be obtained from an effective quality program are:

Improved customer relations
Reduced scrap, rework, and inventory
Reduced costs

Failure Mode Effect Analysis (FMEA)

FMEA is another quality-oriented, statistical process designed to forecast product and process problems so they can be addressed in the design development stage. Many more people may be familiar with FMEA than some of the previous systems because it is frequently a requirement in government contracts, as is VE.

FMEA is a systematic method to define ways in which a design, product, or system can fail, and then it assesses the possible risk of the failure. It also considers the possible cause, effect, security and likelihood of detection. Its forecasts have initiated VE studies toward reducing risk and failure.

Its specific purpose is to assure that the final design does not include features that may lead to a major problem. It should be part of every total process to assure a quality product.

One of the problems that we have found with application of FMEA is taking an assembly or system that is too large for practical evaluation. In one case a complete automobile transmission involved 12 flow sheets, some over 12 feet long. This complexity, combined with limited time for analysis, resulted in dropping the system as being too complex for the operation. It would have been much better to apply the system to operational units which could have been handled much more efficiently and effectively.

Simultaneous Engineering (SE)

"We no longer design the part or assembly and throw it over the wall for someone else to make; we do simultaneous engineering." This statement is being heard more frequently in design operations. What does it mean?

In some cases there may be several multidisciplined product teams made up of engineers, manufacturing, purchasing, and marketing people. The product teams work together during the entire life cycle of the product.

In more formal programs, the product teams report to a steering committee as a liaison group between the team and management. They help to break roadblocks and assure that the overall project is coordinated. The steering committee is usually made up of high-level executives, frequently vice presidents, to see that the job gets done.

Communication is improved through teamwork, and changes are incorporated into the design early enough to have a major effect on the overall cycle time. The result is lower-cost, more reliable products.

SE isn't anything very new. It is the application of ideas that have been discussed for years as good project management practices. If anything is new, it is the high-level steering committee with responsibility for the entire project. This provides the power to break down roadblocks.

SE is a system that is reported to have come to the U.S. from Japan, first implemented by Boeing, then passed on to Ford, and now being accepted by many manufacturers of large, complex products such as automobiles and machine tools.

SE provides total system management from concept to customer. The intention is to develop teams of engineers, manufacturing, marketing, etc., people directly involved in conducting the program. The system is a macroscopic application of teamwork to a project on a broad scale. For example, an SE team may be set up to conduct a total program for the development of a new machine tool or a new automobile. The team brings together representatives of the key organizational elements involved in the development–production process.

Even though a company may be heavily involved in SE, it should be obvious that VE can provide a program to assure development of the best value concept product and process. VE brings together a team of experts to focus precisely on producing the best elements for total value within the overall SE scope. For a skilled value engineer, it should be no problem to set up a program to integrate into the SE program for maximum overall benefit. For example, VE studies of the frame, bed, and other major component systems of a large machine can be conducted during the design phase of the SE program.

The biggest problem is convincing people who are conducting the SE programs that VE can help to achieve their objectives. VE is not replaced by SE. VE can complement SE to make it more effective. One major company has told me that they no longer need VE because they are using SE now and it is more effective. A few questions clearly indicated that VE never had the support to implement ideas that has been given to SE. If it had, SE would never have been needed.

The Theory of Inventive Problem Solving (TRIZ)[13,14]

TRIZ is one of the newer management systems that has been gaining popularity. It was developed in 1940 in the former Soviet Union by G. S. Altshuller when he was employed in the patent department of the Russian navy.

Altshuller believed that most problems engineers face contain key elements that already have been solved. The trick is to find these key elements even though they may be in entirely different industries. Rather than developing as many possible ideas for the potential solution of a problem he felt it was only necessary to identify the potentially best solution. In order to find these key elements Altshuller analyzed over 400,000 worldwide patents to determine their successful key elements.

The evaluation led to his discovery of what he called the Laws of Engineering System Evaluation.

Laws of Engineering System Evaluation

1. Increasing the degree of approaching the imaginary ideal system.
2. Eliminating contradictions.
3. Reducing and increasing the number of subsystems.
4. Increasing the degree of dynamism.
5. Increasing the degree of control.
6. Transition from macro to micro level.
7. Application of different power fields in engineering.
8. Matching and nonmatching characteristics.
9. Removing a human from taking part in an engineering system.
10. S-curve lifeline and three stages of the development of an engineering system.

TRIZ problem-solving tools are based on the Laws of Engineering System Evaluation and include:

Principles and Chart for Eliminating Engineering Contradictions.
The Substance-Field Analysis.
The Standard Solutions for Inventive Problems.
The Algorithm for Inventive Problem Solving.
Special Methods.

The chart for Eliminating Engineering Contradictions identifies more that a thousand kinds of basic engineering contradictions and suggests up to four of the most suitable principles for eliminating them. The Algorithm for Inventive Problem Solving is one of the main problem-solving tools. Use of the Algorithm helps to identify the ideal problem solution. According to the Algorithm, it is not necessary to generate as many ideas as possible and then select the best one; it is enough to reach the best one, the ideal solution.

TRIZ works to identify innovative problems, those that contain a fundamental technical contradiction where improving one quality will adversely affect another. For example, increasing acceleration will adversely affect fuel economy or increasing size will affect weight. TRIZ works to eliminate these contradictions and classifies solutions into five levels.

1. Apparent solutions – No inventions 32%
2. Small improvements of existing products 45%
3. Innovative or essential improvements of existing products 18%
4. New concepts or inventions 4%
5. New systems 1%

The higher level a solution, the more significant improvement of the system.

TRIZ is a highly structured system, and it is cautioned that the system steps must be followed. It is also stated by those skilled in application of the system that substantial education and training are required for successful application.

Software is now available to aid in application of the system, and it is said to eliminate much of the tedium.

Benchmarking

Immediately after WWII many companies were converting from war materiel production to consumer products as fast as they were able to do so. A friend was given the task of converting a recently vacated building to the manufacture of washing machines within six months. At the time the company did not have a product design or the equipment to manufacture them. All they had was the vacant building.

My friend's first step was to buy one of every make of washing machine on the market. He then had long bench-like tables made and had a team of engineers dismantle and lay out the components of every machine on the tables. The next step was to study and select the best components from each machine and incorporate them into a practical operating unit.

He had his clothes washers in the six months but they were obsolete before the first one rolled off the assembly line. However, at the time, they sold every unit they made while they were working frantically to design a new updated machine. This certainly was an example of benchmarking, and it did get the company into production at the earliest possible time.

However, the machines were obsolete by the time the first machine was built because the original manufacturers of the benchmarked products had new ideas they were ready to incorporate into their products as soon as the war effort made it possible. As a result, shortly after my friend's first machines were completed the difference between the competitors' new machines started to show. Of course my friend's company was working on new machines also but they were behind because they did not have the opportunity to develop new concepts based on their past experience. They had to start back with the competition's old designs.

In the automobile industry formal benchmarking programs have been around for over 50 years, but they were usually called competitive analysis. These programs were frequently used as an incentive for developing new ideas. The car is literally cut into little pieces to learn about new materials and processes and the new method competition has developed to regulate windows or control their environment. The system answers the questions, what does competition do to create their structure that makes it lighter or why is their suspension system quieter than ours? Companies use the system to learn how competition does the job. Some small ideas may be picked up and incorporated into next year's models but any major changes may take two years or longer to design, develop, and test. As a result by the time major ideas can be incorporated in the product, competition will have improved on the earlier design.

Relying on benchmarking as a design method leads to a catch-up situation. Benchmarking should be a way to measure your standing vs. competition and to identify shortcomings so you can develop new systems and designs to leap over your competition to become the leader. This requires creativity. Every designer should be familiar with competitive products, their design, operation, and cost. This should provide knowledge and incentive for improvement. However, care must be taken to prevent falling in love with a competitor's design, which can build mental constraints and roadblocks to creativity that are difficult to break.

The main benefit to be gained from benchmarking is determining where you stand in the industry and in what has to be done to improve. Benchmarking should provide the necessary incentive for creative development by identifying areas that must be improved to attain leadership.

Value Analysis/Value Engineering (VE)

Value Analysis/Value Engineering is a complete system designed to clearly define objectives and develop means to achieve them. It has been proven effective worldwide to analyze products and services at any stage in their development, from concept to production. It is especially useful in defining "fuzzy" problems or problems that are not clearly defined.

The basis of VA/VE is an entirely different philosophy from any of the other management systems discussed here. VE concentrates on what a product does for the customer rather than what it is. The term used to define this process is *function analysis*. Function definition is the foundation of VE. It is a technique designed to break down mental constraints to force completely new views of the project. If you are not performing function definition and function analysis, you are not performing VE. In addition to function analysis, a small multidisciplined team of about five people has been found to work the best. The multidisciplined team is required to assure a complete understanding of all facets of the project.

The process is guided by the Job Plan, a simple procedure designed to lead the team through the process. A typical Job Plan is as follows:

Information Phase
Creative Phase
Evaluation Phase
Planning Phase
Reporting Phase
Implementation Phase

The Job Plan separates the necessary project stages clearly and precisely and compensates for the shortcomings of the typical individual. It clearly separates analytic from creative operations to assure maximum creative benefit.

Before starting team actions it is necessary to obtain all available information regarding the project. For a product or process this should include detailed costs, technical drawings, operations sheets, marketing surveys' reliability and warranty information, a set of parts or an assembly, and any other available information. If the project is in the concept stage the information may be sketchy or be only a set of specifications. A societal project may be very fuzzy in that no one has been able to define the project. They may be well aware of the symptoms of the problem but in fact, no one really knows what the problem is. In this case, obtain any information that may be applicable and talk to people. As the process develops, the project becomes more clearly defined and additional information can be obtained that will aid in clearly defining the actual problem.

The first phase of the Job Plan analyzes available information, defines the functions, constructs the ARGUS chart, analyzes function–cost–value information, and identifies the functions offering maximum benefit from creative development. Completion of the Information Phase takes about half the time required to complete a project. However, completion of this first phase is critical to a successful project.

The VE system was invented by L. D. Miles while he was employed by the General Electric Company prior to WWII. At the time, he was working in the Purchasing Department and recognized that it wasn't always possible to obtain critical material. However, when he determined what the material was required to do, its function, he was always able to satisfy the requirement and frequently at lower cost. Following the war he was able to develop the system as it is known today.

VE has been applied to products to improve performance and reduce cost, to create new products, and to manufacturing operations to improve productivity. It has also been applied to improve technical operations from concept to operation of complex equipment. The application of VE to budget and organization analysis has also been effective. In addition, it has been used to analyze jobs, prepare position descriptions, analyze fuzzy community problems such as juvenile delinquency and school–home relationships, as well as reallocating expenditure requirements for more effective operations. In addition, a whole branch of applications has been created in the construction industry to reduce building cost and develop more efficient facilities.

A detailed example of a product analysis is in Chapter 12, Value Engineering — A Total System.

References

1. Odione, G., *Management by Objectives,* Chrysler Canada Workshop, 1976.
2. Drucker, P., Quotable quotes, *Readers Digest,* May 1997, 37.
3. *Coopers and Lybrand Newsletter,* Zero-base approach to management, June 1978.
4. Buik, C. C., The Concept, the process and the value engineer, in *SAVE Proceedings,* Vol. XII, Park, R. J., Ed., Dearborn, MI, 1977, 75.
5. Kepner, C. H., and Trego, B. B., *The Rational Manager,* McGraw Hill, Inc., New York, 1965.
6. Bouey, E. A., Personal communications, 1976.
7. Ealey, L., Taguchi methods, the thought behind the system, *Automotive Industries,* Feb. 1988.

   Miller, R., Continuing the taguchi tradition, *Managing Automation,* Feb. 1988.

   Sullivan, L. P., The power of taguchi methods to impact change in U.S. companies, *Target,* Summer 1987.

   Byrne, D. M., and Taguchi, S., The Taguchi approach to parameter design, *American Supplier Institute, 1986.*

   Continuous Quality Improvement Handbook, Sheller Globe Corporation, Detroit, MI, 1983.
8. *QFD Handbook,* American Supplier Institute, Dearborn, Michigan, 1986.
   Continuous Quality Improvement Handbook, Sheller Globe Corporation, Detroit, MI, 1983.

9. Bradyhouse, R. G., Design for assembly and value engineering helping you design your product for easy assembly, in *SAVE Proceedings,* Vol. XIX, Vogl, O. J., Irving TX, 14.

   Boothroyd G. and Dewhurst, P., *Design For Assembly,* University Of Massachusetts, 1983.

10. Deming, W. E., *Continuous Quality Improvement Handbook,* Sheller Globe Corporation, Detroit, MI, 1983.

11. Akazawa, S., Value engineering, Japanese style, *Value World,* Vol. 4 No. 4, Summer 1981, 3.

12. Main, J., Under the spell of the quality gurus, *Fortune,* Aug. 18, 30 1986.

13. Royzen, Z., Application of TRIZ in value management and quality improvement, in *SAVE Proceedings*, VOL. XXVIII, Vogl, O. J., Northbrook, IL, 95.

14. Braham, J. J., Innovative ideas grow on TRIZ, *Machine Design,* Oct. 7, 56, 1995.

Chapter 3

The Toolbox:
The Application
Benefit Matrix

Every mechanic knows it's better to have the right tool for the job. As a result there are many different styles and types of tools available that are designed to perform different types of jobs. These vary from axes and hammers to pliers and wrenches. Not only do they make the job easier but they make it faster and, in some cases, they even make the job possible. In the past, there were monkey wrenches, smooth-faced adjustable wrenches that wore over time and slipped causing injury to the user. As a result, they became known as knuckle breakers. The need for a better wrench has developed sockets and open-end wrenches that are often designed for special applications and have not only increased safety but made the mechanic's job easier and faster.

So it is with management systems. In the past, systems were developed to do a specific job and, for one reason or another, have lost favor because they were considered to be too complex or simply did not appear to accomplish the results claimed for them. However, just as with a mechanic's tools, one tool cannot reasonably be expected to do all things equally well. Each system has attributes that make its use more successful in different phases of the project development cycle. For example, MBO (Management by Objectives) is applicable to some degree, throughout the entire project cycle but requires specific definition of objectives in the

TOOLBOX

APPLICATIONS

TOOLS	Organization Planning and Developm	Teamwork Development	Operations Simplification	Organization & Operations Cost Imp'vt	Product Cost Improvement	Budget Control	Product Improvement	New Idea Generation	Creativity Development	Product Quality improvement	Manufacturing Operations Improveme	Adm. Operations Improvement	Problem Solving	Information Development	Product Reliability Improvement	Reduce number, Eng. Changes	Software Available	Structured process	Activity Based Process	Function Based process	Total
1 DFA					5	4					5	5	3						5	5	32
2 FMEA										4	5				5				4		18
3 JIT				4	4						5	5							5		23
4 K-T		5			3		4						5					5	5		27
5 KAIZEN		5	4	4	3			5			5	4						4	5		39
6 MBO	5	4	3	5									5						5		27
7 QFD		4			4		5	4		4	3	3	4	5		4		5	5		50
8 Simult. Eng.	3	4			3		3					4							5		22
9 TAGUCHI					4		5			5	5		3		5				5		32
10 Target Costing				5	5						4	4							5		23
11 TQM							5			5	5	5	4		4				5		33
12 TRIZ				4			5	5					4					5	5	4	32
13 VE	5	5	5	5	5		5	5	5	4	5	5	4	4	5	3	4	5		5	84
14 ZBB						4						5							4	5	18
16 Benchmarking				4			4	4				4							5		21
Total	13	27	12	18	44	13	36	23	5	22	46	45	28	9	18	9	8	32	63	10	

Figure 3.1   Toolbox. The matrix shows the effect the tool (Management system) can have on various attributes. The scale reads: 5 Major effect or excellent application; 4 Above average effect but not primary application; 3 Some effect but not good application; 2 Very minor effect; 1 No benefit.

earliest phase. This definition of objectives takes time and has been one of the reasons MBO has fallen out of favor. However, whether it is used or not, it can be beneficial to understand the system so that at least parts of it can be applied when the occasion arises.

ZBB (Zero-Based Budget) is another system that has fallen out of favor because it has been deemed too complex and time consuming. Some companies, however, have found it to be extremely useful in smaller development programs, and in addition, it contains aspects that can be invaluable in applying target cost methods. Some methods such as K–T (Kepner–Trego) are excellent for problem solving; others such as Taguchi Methods are best for manufacturing quality improvement, and still others for simplifying products for manufacture.

All of these tools are dedicated to improving the project based on its appearance or activities involved in its production or application. Only one of the tools is based on changing viewpoint to break down mental constraints and prejudices: that is Value Engineering (VE). The VE system forces people to think differently to develop new products and services based on best value. Because it forces thinking, it can be used in a wide range of applications not necessarily devoted to business or manufacturing.

The Toolbox matrix (Figure 3.1), compares the attributes of each method discussed in the chapter and offers a method for evaluating others as they may be developed. The method used in the comparison is based on a simple five-digit evaluation. If the system is particularly suited to an application, it is rated a 5. If it could be beneficial but not its primary application, it is rated a 4, and a 3 if it has only a minor application. Any rating below a 3 indicates little or no beneficial effect can be expected from the system for this application. For example, DFA (Design for Assembly) is particularly good in simplifying products for manufacturing (5); it does have an effect on controlling cost (4), and some effect on problem solving (3). However, K–T is particularly good at problem solving (5) and, although it can aid in product improvement (4), other systems are better.

The Argus Chart in Figure 3.2 indicates the functions in a product development cycle. This is an abridged version of a chart of a major company cycle which was conducted during an intensive VE project. It is not a flow chart in that it does not follow a before and after sequence. Rather, it is a cause-and-effect diagram based on the Theory of Function Relationships (TFR) which is discussed in detail in Chapter 12, The Argus System. The various tools are shown related to the functions where they can be most beneficial to the system.

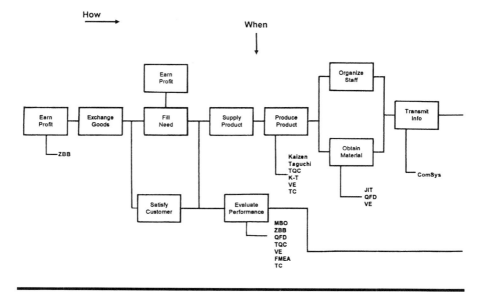

Figure 3.2   Project Development System — Argus Chart

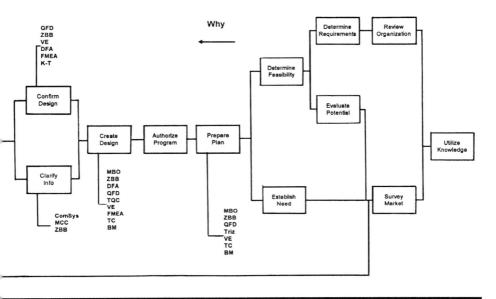

Figure 3.2   (Continued)

THE ECONOMICS OF PROFIT

Chapter 4

Cost and Its Elements: The Driving Force

Introduction

The goal of every company, if it expects to perpetuate its business, should be to earn a profit. In addition, the equivalent of profit should be to operate as efficiently as possible. This means that government as well as business should make every effort to keep their costs under control. However, in order to do this, cost and its effects must be thoroughly understood.

In the first place, it must be clearly recognized that the cost of operating an organization should be less than the amount of money collected or the revenue of the operation. This may sound like a ridiculous statement, but many companies as well as government operations are constantly spending more money than they take in. Even volunteer organizations forget that there must be a balance between income and expenditures. In some cases, this may simply mean that the manpower available to do a job is not adequate to achieve the goal.

Some companies manufacture products and others provide services. However, in either case the cost of operating the business is of key importance. You say everyone knows that. Yes, they do but they seem to forget that to achieve this goal of profit by maintaining a balance of expenditures to income takes effort and know-how.

Experience has shown that even the automotive industry, which exerts a major influence on our country's economic behavior and tremendous influence on many phases of industry, cannot control the prices it pays for basic materials and services for production. It should, therefore, be quite clear that we cannot buy materials for less than the supplier is willing to accept and we cannot sell products for more than the customer is willing to pay. Consequently, only one avenue to increased profit remains, and this is to control cost at every level of the organization. In order to do this successfully, it is necessary to understand cost, its content and effect on every aspect of an operation.

After working many years as an engineer in different types of companies and as a consultant helping companies to improve product performance and cost for profit improvement, I have come to believe that engineers do not understand cost, nor its effect on the product and overall profit. Although in many cases they are expected to work toward a target cost, product performance is their prime objective. Cost is always a secondary goal. This attitude must change. A first step could be to increase emphasis on cost in basic engineering education. This section on cost is designed to create an awareness of the facts of cost and a basic knowledge of how cost can be used for product advantage.

Target Costing

Target costing is so basic that it is hard to believe that it is not a universal practice. However, although many companies do practice target costing, many do not and, in addition, those that do, frequently do not apply adequate discipline to make it work effectively.

The primary consideration in target costing should be to know what the product will cost before it is designed. After the product has been designed and tested it may be too late to get costs in line so that a reasonable profit can be assured. Henry Ford[1] recognized this fact back in the early days of his company as noted in the excerpt from his 1923 book *My Life and Work* in Figure 4.1. Although his method may sound primitive by today's standards, the fact remains that the time to know how much a product will cost is before it is designed.

Target costing requires product teams to meet cost goals that have been established before detailed design has begun. Goals are based on a thorough analysis of customer wants and needs and company profitability requirements.

Finally, the automobile industry is waking up to this fact, and their premise, as stated by one company, is that no longer will the design drive the cost; in the future, the cost will drive the design. This is as it should be. However, to achieve this premise it is necessary to change attitudes

"Our policy is to reduce the price, extend the operations, and improve the article. You will notice that the reduction in price comes first. We have never considered any costs as fixed. Therefore we first reduce the price to the point where we believe more sales will result. Then we go ahead and try to reduce the prices. We do not bother about the costs. The new price forces the cost down. The more usual way is to take the costs and then determine the price, and although that method may be scientific in the narrow sense, it is not scientific in the broad sense, because what earthly use is it to know the cost if it tells you that you cannot manufacture at a price at which the article can be sold? But more to the point is the fact that, although one may calculate what a cost is and of course all of our costs are carefully calculated, no one knows what a cost ought to be. One of the ways of discovering... is to name a price so low as to force everybody in the place to the highest point of efficiency. The low price makes every body dig for profits. We make more discoveries concerning manufacturing and selling under this forced method than any method of leisurely investigation." Henry Ford, *My Life and Work*, Double-day, Page and Company, New York, 1922, pp. 146–157.

Figure 4.1

to create an awareness of cost in every area of the company and to create a disciplined process that will make attainment of the goal as important as designing a product that works at any cost. The premise for the future should be a product that works for the defined cost in the marketplace.

This goal will require a different attitude toward cost estimating and design. Cost estimating must provide reasonably accurate information when it is needed and develop more accurate information as the design progresses. In addition, there must be discipline in the design system to assure that effort will be placed as much on controlling cost as developing a product.

In *When Lean Enterprises Collide*, Robin Cooper[2] allocates a chapter to target costing. He discusses the survival triplet which includes factors that he feels must be considered and their importance to achieving profitability. These factors include price, quality, and performance. He also shows how these elements are basic to target costing. He points out how several Japanese companies apply several cost control methods during the design development cycle to assure that the target cost is achieved. This is the way it should be done. Although many U.S. com-panies complain that it takes too much time and manpower to apply continuous cost reduction operations, the time and manpower are always found to try to reduce the cost after the design is completed. Unfortunately, this approach is usually not very effective.

Cost Creep and Discipline

Cost creep during design and development constantly causes the currently estimated cost to increase. Creep is the slowly increasing product cost resulting from the addition of features, revised specifications, normal economic inflation, and changing market conditions. This creep appears to be inevitable but can be controlled by an integrated, disciplined operation to manage the system cost. During the design development cycle, it is expected that the more that is known about the design the more accurate the estimate should be. However, the counter force of creep is constantly driving the actual cost upward. With a well-constructed cost tracking system, this trend should be visible and steps can be taken to reduce and control the cost to the target.

A well-conducted cost analysis in the early stage of design will identify critical cost areas in the project and should provide recommendations to reduce the estimated cost well below the target cost. However, cost will continually creep upward for various reasons, requiring constant vigilance for control.

Lack of Discipline

It is a fact that most companies do not apply target costing. I have seen products that have been designed and tested and ready for production before it was discovered that the material would cost more than the proposed selling price. If a target cost had been set identifying the basic cost elements, such as material and labor, this problem would have been considered and corrected at an early stage in the design process. Although this is an extreme case, it is not an isolated one.

We were recently involved in a project for an automobile company that was estimated to be substantially over the cost target. The project was in the late stages of product development, and it was feared that management would drop the program if costs were not reduced. We were asked if we could help and proposed a program to solve the problem. A part of the program was awarded to see what would happen. The initial phase was eminently successful. Three workshop projects identified opportunities to improve cost by 20 to 30%, but we could not get people from the company or its suppliers to participate further, despite the results.

They claimed two excuses, the first being that they had already taken all of the cost out of the design and there was absolutely no way any additional cost could be eliminated. The second was that they didn't have the time or manpower to participate.

The product was considered to be so essential to the marketplace that the program went ahead anyway. The product is now in the market and doing poorly because the cost is too high. The solution now is to spend an additional substantial sum of money to make major content changes in the product to better meet competition. However, it will take about two years to accomplish the goal and a major expense in addition to lost revenues.

Different Goals

How could this happen? The company had sophisticated cost tracking and control systems; management was totally committed to target costing, but still the project cost performance was poor. At least part of the answer was that although the top management was committed to a target cost policy, the middle and operating management marched to a different tune. Their objective was to design an operating system or component. If it didn't work by a certain date, they were in trouble. If a cost is too high, the usual process is to get the product into production and reduce the cost later. It never happens! What is more important, we never learn. The same thing happens over and over again.

Cost and Achievement

A serious program to follow up on both department performance and product cost is required. In this day of computers, the cost part is easy. However, the achievement part is difficult. It is possible to meet cost targets without meeting performance goals. For successful target costing, product design targets must be tied to budgeted target costs.

Method Is Required

A cost target is of very little use if the planners and engineers do not have an effective method to help them achieve their target and maintain cost levels. Setting the target is based on economics. Creating a design that meets the cost goal requires flexibility and creativity through a thorough understanding of cost and function. A method is needed to help in this dilemma to force us to see things from a different viewpoint. Although the methods and means to do this have been available for many years, lack of a complete understanding of the process has tended to restrict their use. However, once one has learned the techniques, they become second nature. The process is discussed in detail in Chapter 12, Value Engineering: A Total System.

Where Does All The Money Go?

Whether the company revenue is in the billions or only thousands of dollars why is it that after all of the bills are paid only 2 to 4% is left for profit? Where does all the money go and who is responsible for it? These are difficult questions to answer, not because the data isn't available but because it isn't in a form that the spenders can understand.

There are at least two basic reasons for these questions. The first is that if you don't know where the money goes you can't determine if you are getting good value. The second is that if you don't know where the money is going you cannot take effective steps to improve value and increase profitability.

There are two primary elements to the question. The first is cost. Cost is a fact and is related to the expenditures for products and services. The second is value and is related to the opinion of the buyer or customer as to what the product is worth based on what it does for him.

The owner or manager of a company may not consider the value of his money to be very high if he has to spend thousands or millions of dollars to get 1 or 2% return. He could do better in a bank. However, he could also help his company to do better if he could determine the difference between cost and value for the materials and services he is buying.

Cost has been defined as the expenditure of capital to gain a product or service. This capital may be measured in dollars, effort, time, or a combination of all three. Value, on the other hand, may be a measure of the result of an effort to achieve some accomplishment such as transportation, performance, or in many cases to satisfy an ego need. Value is usually measured in the same units as cost. When this measure is dollars, confusion is often the result.

Companies spend money every day. In almost every case in which companies are producing the same product, some are more profitable than others. In other words, some companies get more value for their money. Why is this? Where should you spend your money to get greater value? To answer this question, we must answer the first, where does all the money go?

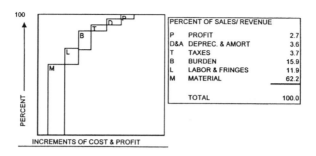

Figure 4.2   Distribution of Expenditures in a Manufacturing Company

After some research I have determined that in manufacturing companies about 60% of all revenue is spent for materials. This may run higher in some companies but not much less in most. The chart in Figure 4.2 indicates the primary elements of cost and their distribution. With such a small amount left for profit, one might ask how can we get a better return on our investment?

In difficult times the classic way to increase profit is to reduce the work force. In the case of the model this means labor and variable burden, or 28% of expenditures. However, in reality only 11.9% can be directly affected. If a third of the work force was eliminated, the profit would be increased by 4 to 6.7%. However, operations would suffer because there is never an effort to redesign the jobs to satisfy the new circumstances. This takes time and an immediate reaction is required. So we try to do the same job in the same way with fewer people.

The better solution is to attack material cost on a regular continuing basis, because it offers a five times greater potential for profit improvement. Just a 3% reduction in the cost of material could increase profit to almost 5% and could be accomplished without major effect on operations.

In the case of one major U.S. corporation, a 3% reduction in material cost would have turned a very major loss into a reasonable profit.

Company management must become aware that there is an opportunity to remove at least one third of the cost from their products. This means an opportunity to increase profit to over 20% of sales or to become more competitive by reducing prices. For maximum benefit this effort must start in design and be carried through to production with discipline and determination.

In addition, the engineer must come to recognize that he has a direct responsibility for company profit. Although he may feel that his primary responsibility is to design an acceptable product, the decisions he makes

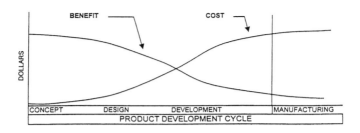

Figure 4.3   Relationship Saving Potential to Time

in the early design concept stage and during the entire product development cycle will have a direct impact on profit. The product size and
complexity affect material, labor, burden, and profit. If the product meets
all product design requirements within the cost parameters established by
finance to assure a profit, he has done his share to contribute to profit.
On the other hand, if the product is over cost targets, the potential profit
will be reduced. If the product complexity introduces additional rework
or excessive warranty cost, his decisions may very well eliminate profit.

Some companies estimate that although the cost to engineer a product
may be about 5% of a product cost the design decisions affect about 80%
of the product price. Others believe the design percent of product cost
can be as high as 5 to 10% of product price and that it is increasing. In
the automobile industry until recently a product could be expected to
have a life expectancy of about 5 years. Now, product lifetimes are getting
down to 2 to 3 years to satisfy changing market conditions. As the lifetime
is becoming shorter, the time to pay off expenses for the new design and
the time to earn profit lessens, and the engineering cost increases as a
percentage of product price.

This situation clearly points out that the earlier in the design stage that
changes are made to control costs the more beneficial they will be. The
chart in Figure 4.3 illustrates this condition.

Cost as a Measurement Tool

We talk about target costing, cost visibility, cost models, etc., but it all
boils down to designing a product for a cost that will allow a price that
will be acceptable to the customer. This cost must also be low enough
to allow for the required expenses in operating the company and still
make a reasonable profit possible. The cost revenue breakdown for a
typical manufacturing company was shown in Figure 4.2.

In most companies the sales or marketing departments will set the price based on their knowledge of the market, benchmarking, focus groups, and other techniques. The accounting department can then establish the target cost for the manufactured product. In our case the manufacturing and design cost will be considered to be the same, which means the engineering department must design a product to meet customer requirements for a target cost that will allow a price that the customer will be willing to pay.

In many, if not most, cases this does not happen. Usually, during the course of the design phase costs creep up because of design difficulties, additional features added by sales, marketing, or a company executive, and in many cases, design considerations to make the product better. There are other pressures that add to the cost, and these include the engineer's priorities, which will be discussed in a later chapter.

Under all this constant pressure the cost moves up, never down, and the original target is forgotten. For this reason constant surveillance to compare cost to the target is required.

Cost Estimates and Time

One of the conditions that make cost reduction programs necessary is poor or inadequate cost estimating. It's imperative to know if a product will be potentially profitable before a large amount of money has been spent on its design and development. This statement is so basic that it might be said that everybody is aware of it; unfortunately, in practice product cost usually isn't defined until a considerable investment is made — money that can never be recovered.

As already pointed out we are aware of cases where products have been designed and developed before it was determined that the material would cost more than the targeted selling price. Unbelievable, you may say. Yes, but it happens. A good idea does not always mean a profitable product.

The dilemma is that the time to determine if the product will be profitable is in the concept stage before the product is designed but when initial sketches have defined a potential product. This means it is important to develop the ability to forecast costs on very little information. This is not as difficult as it may seem. If all of the available information and tools are used, they can produce estimates with an accuracy of ±15%. This variation may be as small as 5% if the product is similar to other products in the line or as great as 20% for an entirely new, complex assembly.

Of course, the more information that is available, the more accurate the estimate should be. However, when all of the information is available

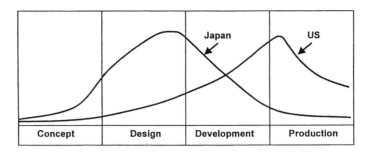

Figure 4.4   Engineering Product Changes vs. Development Cycle Time

to make an accurate estimate, there may be little benefit in learning that the product is way over the target. It may already be too late to prevent major expenditures that can never be recovered. This situation causes, at the very least, the need for a vigorous program to reduce the cost to a profitable level. This can be done, but it would be more practical, productive, and profitable to do the necessary work at an earlier stage.

It is well known that the Japanese take the time to thoroughly plan their project development program before they start major development. One of the advantages of this is that there are far fewer changes during the development cycle, thereby reducing both development time and cost. The chart in Figure 4.4 contrasts the number of product changes in Japanese and U.S. Programs. In the Japanese process, most changes are made in the early design stage. In the U.S. most of the changes are made in the later part of the program, frequently necessitating tool and process changes as well as design changes and often a loss in both time and money.

Cost Projections

The need to know the potential cost of a project is not limited to manufacturing projects. Cost must also be forecasted for major capital investments such as power plants, pump stations, buildings, and even larger projects. In researching this requirement it became apparent that maximum detail does not always produce accurate estimates. It is frequently possible to project the cost at an early stage to the same level of accuracy as a detailed estimate. This is partially because once the targets are set, cost can be controlled to meet the target. Without the target, the cost constantly increases.

A reasonably accurate cost can be determined with practically no detailed information. In order to achieve a result the first and most important technique to use is comparison. Compare the new design to a similar design and identify deviations. The cost of the deviations can be added to or deleted from the base design cost for an initial concept cost. This incremental estimating does not cause the same problem noted in the discussion on ZBB because the incremental estimated cost is based on the cost of the entire project. It can readily be seen how the effect of the incremental cost change will affect the cost of the entire project.

Another useful estimating technique is commonly referred to as Pareto's Law of Maldistribution,[3] which will be referred to in more detail in this section. Pareto's law says that 20% of the parts will control 80% of the cost. If the high-cost parts are determined and evaluated, the major part of the total cost will be identified.

These methods must be used by skilled technicians in addition to a myriad of other tools that many people have developed through practice over a period of time. For example, parameters such as dollars per pound, dollars per cubic foot, per yard, or even per foot of diameter per mile are discussed in cost Analysis Technique 4. These parameters have been used to define concept cost for everything from automotive components to major oil field development projects, office buildings, and power plants, ranging in cost from less than a dollar to several hundred million dollars.

The chart in Figure 4.5 shows estimated product cost during the design and development phase of a project. The chart is based on discussions with several highly skilled specialists in the cost field. They are all engineers and have had many years of experience in design, development, and management. The chart traces the probable accuracy of a cost estimate at the various stages of a project. For example, at the concept stage a cost projection could vary from ±15 to 20%, depending on the complexity and originality of the project. As information becomes available, the margin for error reduces and narrows to about ±2% at the time of release for production.

However, once the estimated cost has been determined and the cost target set, the estimated cost must be constantly monitored to determine its potential cost at that stage so that, if necessary, immediate steps can be taken to bring it within acceptable limits. I am aware of a project that had increased in cost by 300% over the original estimate. This increase was not because the original forecast was wrong but because once the project was approved and under way, additions for desirable features were added during the course of design.

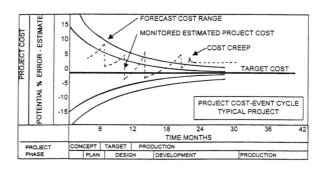

Figure 4.5   Estimated Cost Accuracy During Development

In most cases cost will be over target right up to the time of release for production. This cost over target can come from only one place, potential profit.

Monitoring cost against a target highlights the need for cost control and cost reduction. In many cases it is possible to modify the initial concept to bring the estimated cost in line with forecasts. However, there are only so many times when features can be eliminated or the initial metal gage reduced before the product has lost its potential. In addition, there comes a point when expenditures for tools make changes difficult, and worst of all, there may be a time when so much has been invested that a lower profit must be accepted to remain competitive in the marketplace.

Definitions

Cost is the expenditure of money, time, labor, etc., to obtain a requirement. In business this means the expenditure of all of the necessary components to produce a product or to provide a service. In any organization this means the expenditure of any effort, whether it is paid or volunteered, to achieve a goal.

Cost is a fact. It can be measured and tabulated. However, the process can become so complex that it is frequently forgotten that the objective is to measure performance against a goal.

In business there are many kinds of cost, and many different meanings and uses for each. For the purpose of this text, we will keep the definitions that are necessary to identify and to control costs to a minimum and will, therefore, not guarantee a complete glossary of cost definitions.

When someone tells you how much something costs, the first question should be, "What does that include?"

The reason is that not only are there a number of different cost terms, but frequently these terms may have different meanings and content, even in the same company. The cost terms that will be of the greatest importance to us in this text are defined below.

Design Cost — This cost is the sum of all material, labor, and variable burden required to produce a product. This cost is basically made up of three components, material (M), labor (L), and variable burden (B_V). The variable burden is that part of the burden or overhead necessary to make the part and is discussed later.

$$C_D = M + L + B_V$$

Material — This includes all raw material (steel, aluminum, pipe, tubes, plastic resins etc.), hardware, and purchased items (knobs, fasteners, decals, etc.) necessary to make the product. Some of this material may be charged to indirect accounts and may not always be included in Bills of Material (BM).

Direct material — All raw and purchased material that becomes a part of the end product.

Indirect material — Material that is necessary to the manufacturing process but does not become a part of the product. Lubricants, wiping cloths, gloves, marking equipment, etc., are generally considered to be a part of the variable burden.

Unidentified and other costs — In some cases, there are considerable costs that may not be identified directly with the cost of the product. The cost of painting is one. Painting can be a direct item or part of the burden, depending on company practice. Major rework to bring a product up to standard is another that should also be considered part of product cost. Freight cost incurred in transporting materials to and from a manufacturing site for assembly or other activities can add appreciable cost to the product. Some products may require significant packaging to protect them during shipment or storage. These costs can include boxes, special wrapping, and racks for shipping or storage. In addition, there is always inventory, and excessive inventory cost should be considered as part of product cost. Each of these costs should be identified and included as part of product cost when it is deemed warranted.

There are also other costs that must be identified with purchased items. In some cases, the cost of a purchased item will be shown as material with no labor or burden. In other cases, the raw material may be shown but other costs may include the suppliers' manufacturing costs as well as administrative costs and profit.

When analyzing costs, these sometimes not-so-obvious costs should be considered and identified as a part of the overall cost of the product and identified under material.

Labor — Manpower expended in producing a product or in performing a service. Labor may be direct or indirect.

Direct labor — Labor that can be traced directly to the manufacture of a specific part or labor that changes raw material into a salable part. Examples would be wages paid to the stamping press operator or workers assembling components fabricated for assembly into a system component.

Indirect labor — Labor which is necessary to the manufacturing process but is not directly traceable to a specific part. Examples would be material handling, inspection, receiving, shipping, maintenance, etc.

Burden — Burden, sometimes called overhead, includes all costs incurred by the company that cannot be traced directly to a specific product. Included in burden are management, engineering, purchasing, accounting, real estate, etc. The Accounting Department determines the burden rates based on overall company operations and assigns rates to various operations. Burden rates consist of both fixed and variable categories and separate rates are often established for both. The methods for calculating burden rates vary by company and frequently by industry.

Fixed burden — This includes all continuing costs regardless of the production volume for a given item and includes salaries, heat, light, rent, insurance, and other continuing items.

Variable burden — This includes fringe benefits and any overhead costs directly related to manufacturing the part such as power, expendable materials, sometimes a proportion of the heat, light and space being used in the process. A way to look at variable burden is to consider costs that could be eliminated if the specific job was to be terminated or the cost that would not be incurred if the part was not produced.

Fringe benefits — These costs are usually considered to be a part of variable burden. However, there are some companies that do not separate burden into fixed and variable. In these cases it is important to identify the fringe benefits. Fringe benefits are those costs that are directly related to labor costs and include such items as vacations, holidays, hospitalization, and other benefits. These costs range from about 35 to 75% of labor costs depending upon the company and the industry.

Manufacturing Cost — For our purposes this cost is the same as design cost.

$$C_{MF} = C_D = M + L + B_V$$

Incremental cost — All variable costs do not vary in direct proportion to the increase in the change in the level of activity. Some costs remain the same over a given number of production units, but rise sharply to new plateaus at certain incremental changes. The costs thus affected are incremental costs.

Allowance — Costs other than material, labor, and burden which must be included in the total cost of a product. In some cases, these costs include packaging, scrap, inventory losses, inventory costs, etc.

Administrative Costs — Costs incurred in the administration of the company. These can include research, sales, advertising, etc. They are usually presented as a percentage of sales.

Profit — This is the amount of money retained after all expenses have been paid; it is the earned income. Profit is usually represented as a percentage of sales.

Total Cost — This is the sum of all costs expended by the company and includes material, labor, burden, taxes, general and administrative expenses, and profit.

The chart in Figure 4.2 shows a cost breakdown of all of the major components for an actual manufacturing company. Although the elements of the chart will vary from company to company, several factors appear to be standard for most manufacturing companies. The first is that about 60% of total revenue generally spent on material and direct labor with no burden is about 3% of total revenue. This clearly points out that in order to make major improvements in a company's profitability it is important to consider product improvements that affect material. The best time to make these improvements is in the early design phase of a project.

Standard Cost — This is a theoretical cost based on the weighted average of a number of units. This cost may be generated by purchasing, manufacturing, or accounting to determine performance as measured against a goal. The number does not necessarily apply to a specific item; however, it is often given as a product cost.

Standard costs are an accounting measure to determine performance against a goal such as a budget. They may be used in purchasing, manufacturing, and accounting but are not applicable in design or manufacturing cost analysis. A typical method for calculating standard costs is shown below.

Standard Cost Calculation — Assume that the purchasing department has decided that they will buy a specific material at $0.10 per piece for 100,000 pieces. If they decide to buy 50% of the material from supplier A at $0.10 each, 30% from supplier B at 0.12 each, and 20% from supplier C at $0.14 each, the standard cost for the piece would be $0.1140 each. This cost is arrived at by the following calculation based on 100 pieces:

$$
\begin{aligned}
50 \text{ pcs. } @\$0.10 &= \$5.00 \\
30 \text{ pcs. } @\$0.12 &= \$3.60 \\
20 \text{ pcs. } @\$0.14 &= \$2.80 \\
\text{Total purchase cost} &= \$11.40
\end{aligned}
$$

For a budgeted cost of $0.10 there is a variance of (+)$0.0140 per pc. or $1,400.00 over target for 100,000 pieces.

If the same rationale is used to measure manufacturing performance, the result might look like this based on 100 items:

50 items, material $0.075, labor + variable burden $0.025 = $0.100 cost = $5.00
30 items, material $0.060, labor + variable burden $0.025 = $0.085 cost = $2.55
20 items, material $0.080, labor + variable burden $0.027 = $0.107 cost = $2.14
100 items   total cost: $9.69

Standard (manufacturing) cost $0.0969 ea. For a budgeted cost of $0.10 the variance is −$0.0031 ea. For a production run of 100,000 units the benefit is $310.00 under target.

However, in manufacturing there may be additional costs not included in the standard cost. Among these additional costs could be scrap, rejects, and expendable materials used in the process such as cutting oil. These costs may or may not be included in the standard cost, depending on the

Cost Questionnaire

1. The normal lag in actual cost figures in a typical system is not over 20 days.	T F
2. Actual material costs in a standard cost system are usually reported by part.	T F
3. Warranty expense is regularly reported in manufacturing burden.	T F
4. Unused capacity costs are excluded from cost accounting product cost figures.	T F
5. Freight and duty costs are normally charged directly to the part involved.	T F
6. Normally, scrap material costs are applied only to the part creating the scrap.	T F
7. Direct labor costs as identified in the cost system normally include fringe benefits.	T F
8. Normally when offal is used to make a part, the direct material cost is identified as "no cost."	T F
9. Burden rates are usually adjusted to reflect the "actual" volumes.	T F
10. Scrap costs are reported in product costs.	T F

Figure 4.6

standards and procedures of the company. In many cases they are in separate expense records.

From these examples it should be obvious that when given the standard cost for a product it is important to ask what is included in the cost and how was it calculated. In addition it should also be apparent that the standard cost does not apply to a specific part only to a group or family of parts; for example, parts purchased from a particular supplier.

The Cost Questionnaire — The cost in Figure 4.6 was created by the division controller of a major manufacturing company. The object was to illustrate that there is no firm or fixed meaning to many cost terms; therefore, there are no correct answers to the questionnaire.

Sources of Cost Information

Cost visibility begins with an analysis of total cost, progresses through an analysis of the cost elements, and finally analyzes component and/or process costs. For a successful analysis, the best cost information available is required. Although this information is usually available, it is not always in the same place. Sources of this information are to be found in many areas of a company such as:

Accounting — Current and historical costs, these are usually total costs.

Purchasing — The cost of materials, tools and fixtures, material weights, freight, and all other purchased components. In some cases various labor contract costs are also to be found in purchasing. In addition, many large companies maintain cost estimating departments that can provide estimated product costs.

Cost Estimating and Planning Operations — Estimated costs of parts and tools, processes, and estimated weights. These costs can be forecasted costs based on marginal information or estimated costs based on design drawings as the design progresses toward completion.

Suppliers — Estimated costs and/or process information and material prices. Working more closely with suppliers earlier in the design phase of a project than has been the practice in the past, not only makes it possible to obtain additional cost information, but the supplier can frequently offer opportunities for improving designs or processes.

Feasibility and Guidance Estimates — Some companies are moving cost information earlier in the development cycle by creating the ability to estimate cost directions based on design characteristics and "rule of thumb" data. As the design becomes more clearly defined, the ability to make more accurate estimates becomes greater.

Cost Visibility Techniques

Pareto's Law of Maldistribution — Vilfred Pareto (1846–1923), an Italian political economist, in his studies of old and new societies, observed a common tendency of wealth and power to be unequally distributed. He came to believe that there is an obvious inequality in human endowment and in every society there is, and must be, a minority that rules the majority. Because of its wide occurrence he elevated this phenomenon to a universal law. These ideas no longer fit today's democratic thinking, which stresses the underlying equality of all men. However, this is not true for his distribution laws.

It has been known for many years that costs tend to group in accordance with this Pareto Law. This relationship has come to be known as the 80/20 law. However, other people have redefined this law as follows: "In any series of elements to be controlled, a selected small fraction in

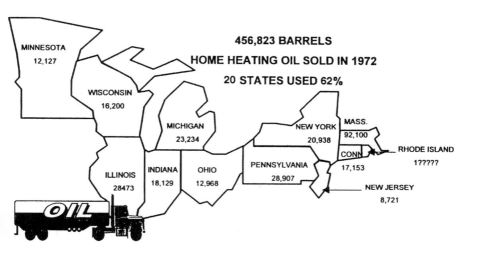

456,823 BARRELS

HOME HEATING OIL SOLD IN 1972

20 STATES USED 62%

Figure 4.7   Cost Distribution — Gas Turbine Engines

terms of number of elements usually accounts for a large fraction in terms of effect;" or in much simpler terms it says that in most distributions, more often than not, 80% of the value is invested in less than 20% of the items. In this form it has popularly become known as Pareto's Law of Maldistribution. In this form it is one of the most valuable tools for cost visibility and estimation available today.

The distribution works. It has been applied from jet aircraft engines to pipe couplings, automobiles and their components to electronic instruments, buildings and structures, and in every case the ratio works out to within a few percentage points. In addition, budgets, procedures, and various other practices also meet the magic ratio.

To show the diverse application of the Pareto rule, the chart/map in Figure 4.7 shows that about 62% of all of the heating oil sold in the U.S. is consumed in about 20% of the states. The chart in Figure 4.8 indicates the cost of the main components of two small gas turbine aircraft engines. Although they are not the same parts in each engine they represent about 20% of the total number of parts in the engines. They also represent about 78% of the total cost of each engine.

Knowing that such a ratio exists and understanding how to use it can be invaluable in estimating any type of effort. For example, an estimate of the cost of the main components can be made in the early stage of a design. In all probability these will number only about 20% of the total number of components, but they will represent about 80% of the total

Model 421	$	Model 456	$	Model 217	$
Rotor Assembly	2706	Fuel control	12000	Fuel Control and Pump	3450
Fuel contr. & Pump	2280	Eng. Assy & Test	6000	Eng Assy.	1400
Rotor Assy. & Fan	2041	Free Turbine Rotor	3200	1st Stage Axial	820
Eng. Assy.	1443	1st Stage Axial	3000	1st Stage Turb. Rotor Assy	682
Nozzle Assy. Free Turb.	1293	2nd Stage Axial	3000	Free Turb. Rotor Assy.	680
Turb Rotor Assy.	1279	3rd Stage Axial	3000	2nd Stage Turb. Rotor Assy.	680
Rotor Radial Compr.	842	5th Stage Axial	3000	Starter-Generator	600
1st Stg. Nozz & Comb. Assy	716	1st stg. Nozz & Comb Assy.	3000	Rad. & Axial Diff. Assy.	500
Bevel Gear access. Drive	658	2nd Stg. Turb Rotor & Shaft	3000	1st Stage Nozz. & Inner Comb.	489
Rear Comp. Housing Assy.	557	1st Stage Turbine Rotor	2700	Rotor Inducer	470
Turbine housing	439	Compressor Rotor	2500	free Turb. Nozz. & Bearg. Hsg.	460
Oil Pump	408	Free Turbine Nozzle	2100	Accessory Case	415
Outer Combuster	385	Radial Combuster Hsg.	1800	2nd Stage Axial Rotor	410
Oil Tank & Cooler	370	Governers (20	1800	Housing - Air duct	407
Rotor - Inducer	361	Fuel Pump	1550	Rotor - Radial compressor	300
Segament - Axial Diffuser	325	Accessory Case	1500	Oil Pump	300
Diff. & Comp. Cover Assy.	236	Axial Diffuser & Cover	1500	Housing - Radial Compressor	260
Manifold - Fuel Delivery	214	2nd Stage nozzle	1500	Turbine Housing	240
Oil & Fuel Lines	200	Combuster Basket Assembly	1000	Outer Combuster	230
Ball Brg (2) Free Turbine Rotor	180	2nd Stage axial Stator	1000	Shaft Free Turbine	185
Total	16,933		58,150		12,940
Engine Cost	21,600		74,300		17,660
Percent of Total Engine Cost	78.0		78.2		76.0

Figure 4.8   Heating Oil Consumption in the U.S.

design cost. This will indicate whether or not the product concept is practical from a cost standpoint.

Basic Cost Analysis Techniques (BCAT)

Although the following series of techniques do not have to be applied in chronological order, it is usually best when evaluating the cost of a product or service to start with the manufacturing cost. This cost includes all of the cost elements that are affected in the manufacturing process.

A sample cost chart of an actual automobile door latch is shown in Figure 4.9, and segments of the chart are shown below.

Technique 1 — Determine Manufacturing Cost

First look at the total manufacturing cost and based on judgment, experience, or other cost data that may be available, decide if the cost is reasonable and in line with typical costs for the product or if the product is worth its cost.

Determine Manufacturing Cost
$1.1857

This technique is simple and obvious, but it shouldn't be overlooked as an indictor of where the high cost may be. If after comparison to other similar items Technique 1 tells us that the cost is too high, it is necessary to look at the components that make up the assembly to determine why the total cost is high.

Technique 2 — Determine High Cost Element

In the example, the total component cost is broken down into material, labor, and variable burden. Look at the elements of cost to determine if they are in the proper ratio for this type of product. There should be a normal distribution of costs. Accounting can usually determine the normal distribution of cost in material, labor, and variable burden for a specific department or profit center.

Determine Cost Elements		
Material $0.4751	Labor $0.2802	Variable Burden $0.4304

COST VISIBILITY

Team No._____ Assembly - Part Name ___Door Latch___

Date_____ Assembly Part No. _____

Determine Manufacturing Costs		Determine Cost Elements	
$ _1.1857_ Material $ _0.4751_		Labor $ _0.2802_ Burden $ _0.4304_	

Item No	Req.	Part Name	Raw Material	Labor $	Burden $	Other $	Total Component $	Cost Per Unit $$
1		prim. spring	0.0194	0.0007	0.0026	———	0.0227	
2		prim. spring.	0.0051	0.0010	0.0031	———	0.0092	
3		spring	0.0010	0.0004	0.0011	———	0.0025	
4		lever	0.0082	0.0044	0.0077	———	0.0203	
5		ratchet spring	0.0128	0.0016	0.0049	———	0.0193	
6		reinforcement	0.0048	0.0039	0.0064	———	0.0151	
7		bearing	0.0044	0.0039	0.0064	———	0.0147	
8		clip	0.0029	———	0.0003	———	0.0032	
9		spacer	0.0055	———	———	———	0.0055	
10		ball-end	0.0065	———	———	———	0.0065	
11		spring	0.0154	———	———	———	0.0154	
12		ratchet spring	0.0113	0.0021	0.0056	———	0.0190	
13		ratchet-prim.	0.0442	0.0069	0.0122	———	0.0633	
14		rivet	0.0080	———	———	———	0.0080	
15		activator	0.0128	0.0120	0.0192	———	0.0440	
16		pawl	0.0241	0.0054	0.0088	———	0.0383	
17		pawl (2)	0.0172	0.0019	0.0052	———	0.0243	
18		bumper	0.0040	———	———	———	0.0040	
19		reinforcement	0.0178	0.0015	0.0042	———	0.0235	
20		wedge	0.0180	0.0059	0.0165	———	0.0404	
21		slide	0.0083	0.0013	0.0034	———	0.0130	
22		plate	0.0635	0.0129	0.0224	———	0.0988	
23		act. slide	0.0072	0.0012	0.0032	———	0.0116	
24		lever	0.0089	0.0012	0.0033	———	0.0134	
25		clip rod	0.0052	———	———	———	0.0052	
							TOTAL COST	1.1857	

Notes

P.F. costs are estimated
Material + Labor + Burden = Mfg. Cost
Mfg. Cost + other = PF cost or other Cost

Figure 4.9   Cost Visibility

Compare material cost content to the labor cost content and relate to the normal cost distribution for the operation. Every part in the product can then be compared to the distribution cost to determine if the cost elements are high or low. Again comparison is the major tool.

A cost breakdown may show that $10.00 worth of material and $0.10 worth of labor are being expended on a certain part. If this is the case, it should be asked if we are in business to spend $0.10 on labor for $10.00 worth of material. Perhaps the material supplier should be asked to perform the necessary operation. This could eliminate the labor which may be used more productively elsewhere.

Conversely, it may be found that $0.10 worth of material requires $10.00 worth of labor. If this is the case, the overhead should be broken down into variances, set-ups, tooling, direct and indirect labor, etc., to determine the cause of the high cost. The manufacturing area should be questioned about methods and processes, profit centers being used, burden, capital equipment, labor grades, etc.

Technique 3 — Determine Component or Process Costs

The third technique goes one step further in breaking down material, labor, and burden. To determine the total assembly and component costs as they occur in the manufacture of a part, break down each part as shown in Item 1.

Item No.	Req	Part Name	Raw Material $	Labor $	Burden $	Other $	Total Component	Cost Per Unit $
1		prim. spring	.0194	.0007	.0026	—	.0227	

Item 1 shows the components broken down into elements. Examination of the cost elements indicates each entry to be reasonable until we examine Item 22. Herein lies an area which indicates the highest of all labor costs.

22		Plate	0.0635	0.0129	0.024	0.0988		

Circle this amount and examine it in detail. Determine why this cost is so far out of line with the other components.

This technique gives a very precise and accurate cost visualization. It shows where high costs are being created on a component and element cost basis.

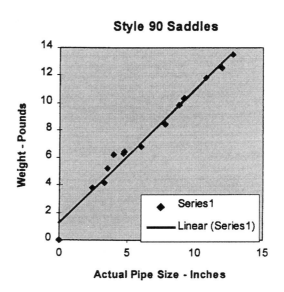

Figure 4.10   Size vs. Weight

Almost every analysis should include the use of techniques 1, 2, and 3
Now think of the third technique in more depth. If we study technique 3 in depth it will be seen that it can be used to analyze parts being assembled into a major sub assembly, major sub assemblies being assembled into a final assembly, and a number of final assemblies being assembled into a final product. Good judgment must be used in the application of this technique and it will also dictate the way the techniques should be used.

Technique 4 — Determine Quantitative Costs

This technique measures cost on the basis of some measurable unit such as time, weight, size, area, etc., and then makes a comparison with the cost per unit of a known good value. It is sometimes surprising how seemingly complex products will fall into a pattern. Two simple product line charts are shown in Figures 4.10 and 4.11. The first chart shows the weight vs. size for pipe saddles based on actual pipe sizes. The second shows the cost for a series of pipe clamps.

Size vs. Weight — Figure 4.10 shows that the weight of each size fitting in the product line falls on a straight line. However, the 4-inch size is substantially above the line. An investigation of this fitting showed that

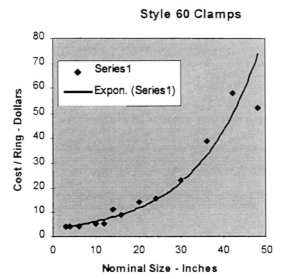

Figure 4.11 Cost vs. Size

additional iron was added to the product in the past to correct a deficiency. This resulted in a high-cost, low-profit part which was identified for product improvement.

Size vs. Cost — The pipeline clamps shown in Figure 4.11 are plotted by weight vs. cost in dollars. In this case each fitting falls on a curve with the exception of the 16- and 42-inch sizes. Again an examination of the products indicated that a redesign was warranted.

In both cases redesign improved the product from a cost and performance standpoint.

Cost vs. Time — This measure can often be used in measuring high-volume production. It can also be used to describe cost per similar product class. Measuring the number of units produced in a specific period of time such as a minute, hour, day, or some other unit can be compared to a similar unit. For example, this measure could run from a number of fasteners per minute to the number of complete automobiles per shift.

Cost per Dimension — Some simple examples of this criteria would be cost per unit of length for an extrusion, cost per unit of volume for a tank, cost per square foot covered for paint. These rule-of-thumb measurements are very useful for estimating requirements in early design stages and can be quite accurate, often to within 5% of requirements.

These cost analysis techniques are listed in Figure 4.12.

Cost Analysis Techniques

---

1. DETERMINE MANUFACTURING COST
   $C_M = M + L + B_{FV}$
2. DETERMINE COST ELEMENTS
   M, L, B_V
3. DETERMINE COMPONENT OR PROCESS COSTS
4. DETERMINE QUANTITATIVE COSTS
   COST/#   COST/Measurable unit
5. DETERMINE COST PER FUNCTIONAL PROPERTY
   COST PER FUNCTIONAL UNIT

Figure 4.12

Dimensional estimates are also used for very complex estimates. For example, the cost of oil refineries, pump stations, pipelines, buildings, engines, and a myriad of other complex products and services are often estimated based on dimensional properties based on long-term experience. Oil refineries might be estimated based on barrels per day, power plants on kilowatt hours, pump stations on horsepower or gallons per day, and pipelines on cost per mile per inch of diameter. In every case there is a set of criteria that can be used to determine cost for a required effort. In many cases the trick is to find the relationship.

Cost per Functional Property — This measure is somewhat different than the other measures identified in this section. In this case the cost is based on what the product does, the functional property. For example, the cost of a wiring harness could be based on the amperes conducted, a mechanical component on the torque transmitted, and a structural member on the weight supported. Again, the measure is a comparison with some other similar unit that is known to be acceptable. This measure will be explored in detail in Chapter 6.

The use of these cost analysis techniques literally explodes costs in such a way that a circle can be drawn around the areas that show where work is required. The functional approach techniques can be used to study the high cost area. It does not follow automatically that high cost is unnecessary cost. It may be, but other tools must be used to find out whether it is unnecessary.

Technique 5 — Determine Functional Area Cost

One purpose of this technique is to help answer the question, "where should effort be applied?" If the study item is part of a simple assembly (2 or 3 parts) the scope is already defined. If the project is a complex

Part name *Microswitch*

Functional Area	Present Cost	Hi	Low
Mechanical	0.20		x
Electrical	0.10		x
Housing and Cover	0.30	x	
Assembly and Labor	0.15		x
Total Cost	**$0.75**		

Figure 4.13   Function Cost

assembly which could have its principle of operation changed by a new design concept, questions such as available time, savings potential, type of improvements, stage of product maturity, etc., should be considered.

Divide the present cost into functional areas to define the project scope. Division of cost into functional areas will pinpoint high cost differently than the usual cost visibility analysis and will help to broaden or narrow the scope of the study. This will direct effort to more profitable areas. An example is shown in Figure 4.13.

Project Justification

In any cost reduction procedure it is important to carefully check the cost benefit. For example, it is important to determine if the potential recommended change will meet acceptable criteria for return on investment. In other words how long will it take to get my money back, and will the project return meet company investment criteria?

Almost any change will require some expense to implement, even if it only involves changing drawings and specifications. If the change will require a substantial expenditure for tools and equipment, the cost benefit analysis may become critical. In most cases the critical part of the estimated saving benefit will be directly related to the cost of labor and burden. The estimated result will be based on $M + L + B_V$, that part of the cost that will be eliminated when the change is implemented.

This is why the use of only variable burden (B_V) is important. Full burden will not be eliminated. Management, buildings, and most other burden expenses will still remain. The only burden costs that will be eliminated will be those directly related to the change. This should include power and light to run the operation and fringe benefits which are usually included in B_V. If it is not possible to obtain a B_V breakdown, then only the fringe benefits should be used. This is to ensure that the estimated

$$R = \frac{E}{B} \times 12$$

Expense E   = $150,000.00/yr.
Volume   = 500,000 units
Piece cost saving   = $.50/Pc.
Potential benefit B   = $250,000.00
Payback R   = 300,000 units or 7.2 Months

Figure 4.14   Payback Analysis

potential saving will actually be removed when the change is made. The estimate will be conservative and assure a benefit.

The expense must include all costs required to make the recommended change. This should include the cost to change drawings and specifications as well as buildings, machines, equipment, and all other costs that may be related to the change. However, only costs related to the specific recommended change should be included. Do not, under any circumstances, include buildings, equipment, or any other wants that are not part of the recommendation. In many cases this practice has caused a very profitable improvement to be lost because the increased cost created a poor payout.

Method 1 — Payback Analysis

To analyze the potential benefit from a change we have always used the simple equation based on two factors shown in Figure 4.14. These factors are the expense to make the change (E) and the potential benefit (B) is the gross saving on an annual or some specified time basis. The expense to make the change would include engineering design and development expense, tools, fixtures, machinery, and any other costs necessary to implement the change. The gross saving (B) is simply the estimated piece cost saving times the annual volume. The payback is (R), the expense is (E), and the benefit is (B). This is multiplied by 12 to get the result in months. This is the time it would take to return the cost of making the change.

Method 2 — Break Even Analysis

The break even chart can also be used to make a comparison between alternatives to determine which may have the better payback. The example in Figure 4.15 shows the result of a comparison between an existing product and operation and a recommended change.

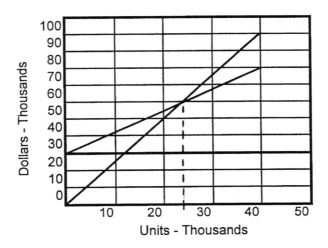

Figure 4.15   Break Even Analysis — Two Alternatives

A. Recommended change
    Volume   V = 40,000 units/yr.
    Expense to change   E = $30,000.00
    Piece Cost $1.25 Per pc.
    Cost to manufacture per year $50,000.00

B. Present condition
    Volume   V = 40,000 units/yr.
    Expense to change   E = $0
    Piece cost $2.50 Per pc.
    Cost to manufacture per year $100,000.00

From the graphical analysis it can be seen that the change will begin to pay back after 24,000 units or about 7.2 months.

Cost Visibility, Budget Analysis, and Fund Allocation

Budget Analysis and Fund Allocation

One of the main reasons for applying any of the management methods listed in Chapter 2 is to improve operations or products in some way. It may be to increase productivity, improve product operation, reduce the cost of products or operations, or to reallocate funds to increase efficiency.

In many cases, if not most, there may be a shortage of funds to carry out a desired program while other operations are busy conducting operations that either duplicate effort or are not very important to the success of the company.

How can the important operations be identified so that the manpower and funds can be redistributed to attain a more efficient, profitable operation? This is usually not an easy task; however, it can be done and the overall organization will be better as a result. Frequently, people recognize that the job they are doing may not be important or necessary but they do not say anything because they are afraid of the consequences. The engineering department in Chapter 12, Value Engineering, is one example; another is an insurance company that is now doing substantially more work than they have ever done before with about half the people. Another case is a government office that reallocated funds by identifying poor value functions, redesigning jobs, setting up new specifications for meetings and a number of other conditions that made it possible to increase the number of client contacts by one third at no additional cost.

Whether it's reallocating the costs in a product for a lower-cost, more efficient product, redistributing the sales space in a department store, redistributing funds in a budget, or analyzing the operations in a manufacturing facility, the process is the same. However, the budget process must start at a very early stage and combine some of the aspects of a product analysis as well as the zero based budget process.

At one time the company that I was working for introduced a new budgeting system to attack and eliminate some of these problems. The system was called PROBE and created a three-dimensional type of budget. As it turned out the project was not very successful, partly because key executives and managers were not properly trained in the benefits of the system and partly because the project champion was not given the proper authority to foster implementation and feedback.

However, the idea of a three-dimensional budget was appealing to me. So I studied the method and realized that although the third dimension was called "function" it was actually another way of describing the activities such as engineering, manufacturing, finance, etc. To my understanding these are really activities; the function is the output of these activities, such as designs, reports, etc.

As I understand function, engineering, manufacturing, finance, etc., are activities that produce functions. The function is the result of an activity and is something we want or need. For example, one of the functions of engineering is to create a design; another might be to confirm the design to assure its operation. Several departments might produce part or all of

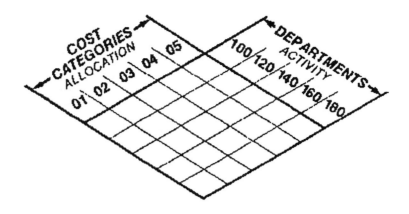

Cost Categories - Departments

Allocations - Activities

Figure 4.16   Two-Dimensional — Activities Budget

the same function. The trick is to properly define the functions. This is the foundation of the budget.

How to define functions is described in detail in Chapter 5, Function, and an example of the result of the system is the engineering department example in Chapter 12.

The key to these beneficial results in a 3D budget is function, just as it has been for all of the other projects and examples illustrated in Chapter 12. An abstract example of how the process works will show diagrammatically how the process identifies the Targets for Opportunity (TFO).

Activities Budget

Figure 4.16 shows the distribution of funds in a typical budget for an organization. It shows that each department allocates funds to different cost centers. For example, depending on the operation, the cost category might be personnel, client research, mail distribution, product design, product testing, packaging, printing, or similar activities.

A typical budget based on departments and its activities is a two-dimensional budget. I call this an Activity Based Budget. In this budget the departments are 100, 120, 140, 160, and 180, and the cost categories established by the accounting department are 01, 02, 03, 04, and 05. In

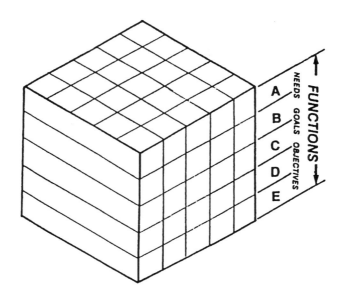

Cost Categories - Departments - Functions

Allocations - Activities - Needs

Figure 4.17   Three-Dimensional — Performance Budget

this budget if we want to find out how much was spent on printing last year we can check category 01 and find the total amount spent in the company and who spent it.

However, if we add a third dimension to the budget called function, the budget becomes a cube like the illustration in Figure 4.17 and interesting new cost visibility becomes immediately available. For successful application of the system, it is important that functions be properly defined. It is not correct to call these functions accounting, engineering, manufacturing, etc. It is important that they be defined as needs, wants, goals, and objectives as discussed in Chapter 5. Remember, the functions are what we pay for. What does the accounting department do? Maintains records, council management, collects data, etc. These are some of the functions of an accounting department. A manufacturing department may convert material, consume energy, receive material; an engineering department creates design, issues instructions, confirms product. How much money is spent on these functions?

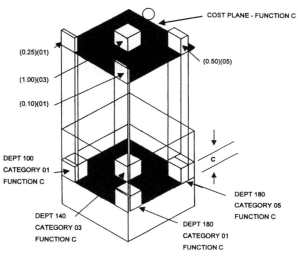

Figure 4.18   Function Cost

The first step is to create an Argus Chart that will define the required functions. The next step is to find out how much the functions are costing. The third step is to determine if they are worth their cost and develop new ways to perform poor value functions.

1. Create Argus Chart — Develop Functions
2. Determine Function Cost
3. Determine Function Value

The illustration in Figure 4.18 shows how the cost of function C might look. In the illustration, Department 100 spends 25% of the cost of Category 01; Department 140 spends 100% of category 03; Department 180 spends 50% of category 05 and 10% of category 01. The total cost of Function C would therefore be the sum of the expenditures from all three departments and their total expenditures in all categories.

Function costs group in strange ways. Just as *confirm design* in the engineering department (for example, see Chapter 14, Figure 14.14) shows a 43% expenditure of funds, *develop client data* may also indicate an inordinate expenditure and indicate a TFO to reallocate resources for a more efficient use of funds. This grouping of costs tends to make value setting possible and shows how to redistribute cost for a more equitable or effective operation.

Summary

These cost analysis techniques are important and can be used in analyzing any product or service to isolate high or unnecessary cost. However, do not overlook the use of the very same techniques to prevent the build up of unnecessary cost in products and services. The item Cost and Inflation in the Appendix gives a good example of how costs have risen over the years. Learning how to isolate high cost is a matter of practice. The more often these techniques are practiced the greater the awareness to cost will become, and the result will be better, lower-cost products and services.

After working with many companies and engineers and managers at various levels including vice presidents, it appears to us that the comments made here apply universally. This appears to be a reasonable assumption because as far as we know no cost or cost analysis course is offered in any college curriculum. It follows, therefore, that the information contained herein must be learned on the job, and if a company is not practicing good cost control and cost discipline, the methods may never be learned. It is important that this section be thoroughly understood before attempting a workshop as described in Chapter 12.

References

1. Ford, H., *My Life and Work*, Doubleday, Page and Company, New York, NY, 1922.
2. Cooper, R., *When Lean Enterprises Collide,* Harvard Business School Press, Cambridge, MA, 1995.
3. Van Den Breckel, A., Pareto and Good VE, *SAVE Proceedings*, Vol. VI, Kaufman, J. J., Chicago, IL, 1971, 287.

Chapter 5

Function: The Foundation of Clarity[1]

Introduction

Look at one thing and see another. See what everyone has seen, but think what nobody has thought. Work as a total person with subconscious and unconscious levels acting in unity with the conscious. Free our minds from the fears of reality. Develop the ability to understand the obvious.

These are expressions on creativity by creative people. The consensus seems to be that to be creative one must be able to see beyond the conscious, beyond the existing.

What is the ingredient that some people have that enables them to break the barriers to visualization, to be able to look at something and immediately think of new exciting possibilities for new products, services, methods, or other useful or satisfying subjects?

This is a provocative question that produces many and varied opinions but no clear-cut formula for producing creative people. It is known that a creative person is somewhat different. It is also known that a creative person exhibits certain characteristics. However, given the same characteristics, another person may not prove to be creative. Many people feel that the seeds for creativity exist in every person. If this is true, it would be exciting to discover the means to release these seeds and to foster their growth for the benefit of mankind.

In all probability, at least one of the ingredients of creative people is the ability to visualize, to detach themselves from reality, to see beyond the stated problem, the object, or the material facts. A creative person must be able to create concepts, broaden and develop them, and out of it all select a new idea, a new approach, or a new solution to a requirement or problem.

Can this be the key that made Kettering, Einstein, Watt, Galileo, and hundreds of others able to produce so many useful ideas and creations? What if everyone could produce results like them?

Function Is the Key

There is a method that offers a key to open the door to creativity. The key is function analysis, and it is the main element in the Value Engineering (VE) management system. Function definition and function analysis are the main elements that set VE apart from all other management systems.

Value Engineering (VE) was originally called Value Analysis (VA), but as skill in its use and understanding of the method grew, the application of the method broadened. As a result, it has become known by several other names which tend to describe the scope of application from product and process analysis to planning and management. In addition to VA and VE, Value Control and Value Management are frequently used names. We have elected to use the term Value Engineering because it is universally recognized around the world. In addition, the method of application is the same no matter how the system is being applied. However, it must be recognized that if function definition is not part of the system it cannot be called VE or any of the other terms used to describe the system.

VE is the result of observations made by Lawrence D. Miles[2] during World War II. Mr. Miles was a purchasing agent at the General Electric Company, and over a period of time he began to recognize that although he was not always able to provide the requested material because of critical shortages, he was always able to satisfy the need with substitutes that satisfied the requirement, the function. In many cases these substitutes not only worked but improved the product and many times, at lower cost. Mr. Miles reasoned that if it was possible to satisfy a requirement at lower cost when the requested material was not available, it should be possible to develop a workable method for creating a lower cost substitute as a standard practice.

The application of the VE system is described in detail in Chapter 12 by application to an actual project. Although the project used in the example is relatively simple, it follows all of the steps in the process.

Chapter 14 discusses a number of other successful projects and ranges from product and process analysis to more fuzzy subjects to show the complete scope of the method.

It has been necessary to separate this discussion of function from the process described in Chapter 12 because in preparation of the text it was found that an understanding of the term function at this stage was necessary for an understanding of several other subjects in the text.

In his position at General Electric Mr. Miles came to realize that function was the key to cost improvement. In a discussion with long-time friend and well-known early VE pioneer Carlos Fallon, Miles said, "Ah functions! We buy only functions. What things do. It is the difference between a word and an idea. However, a Value Analyzed function must offer a better way to do something."[1] He came to realize that we frequently become constrained in our thinking and need a way to help us to open our minds to a wider range of alternatives. As a result, Miles came to two conclusions:

1. Creative thinking is constrained by the physical shape or concept of existing products or services.
2. Concentrating on the need or the requirement, which Mr. Miles called the function, helps to break down the constraints to visualization and offers outstanding opportunities for creativity.

He felt that the conventional approach to product and process improvement to try to make the existing product work better, cost less, or meet some other objective, stifled creativity. The problem is that the way something looks or works tends to constrain creative thinking.

The concept to break this constraint is function analysis. It is disarmingly simple. It is easy to understand without learning complex systems or studying complex technology. However, the ability to effectively use the system requires a thorough understanding of the principles and the determination and discipline to use them.

Function definition is a basic requirement in helping to break down our mental barriers that have developed over a long period of time, in fact our lifetimes. Unfortunately, thinking function and defining functions does not come naturally to most of us. However, function definition and function analysis provide a major discipline for helping a person or group of persons to visualize beyond their normally accepted standards. In fact, the forcing and struggle necessary to properly define functions make it possible to look at what you have seen many times before but to see new and different things, to see your problem in a new light. It can help to achieve the ability to see beyond the stated problem as outstanding people have been able to do throughout the ages. Defining function creates awareness of the project at hand, but first we have to learn to define functions, then we have to learn to define functions that will offer creative opportunities.

On a second assignment for a client I was introduced to the workshop group by the Vice President of Manufacturing who had taken part in the first workshop. In his introduction he pointed out that as a result of the first program, new ideas made it possible to pay my fee in three days. When asked how he was able to do that he simply said, "I walked out into the shop." In looking at this thoroughly familiar world, he saw things that had become commonplace but with a new insight that was worth thousands of dollars.

This is a clear example as to the potential benefit that can be obtained by learning to think function. It provides the opportunity for a person to break his own mental barriers to seeing new things, to eliminate prejudices and provide insights never before possible.

Function Defined

The usual definition of function is the properties that make something work or sell. Miles defined function as a want to satisfy a requirement. Function is the result desired by the customer. Function is what is paid for. Function is a requirement, a goal, or an objective. Function is what the product or service does for the customer; it moves weight, increases pleasure, or some other practical or psychological want or need.

Webster[3] defines objective as an aim or end of an action; it is, as Webster says, the result of an action. In the analysis of a product or service, first the total assembly or operation, then each part or action must be evaluated to determine the function or functions it performs. In order to aid in understanding what is happening, these functions are defined in two words, a verb and a noun. Only when function is defined in terms of two words will the goal of providing a key for maximum creative opportunities be achieved.

In some cases there is a feeling that it is unnecessary to struggle for two-word definitions. It is frequently stated that simple terse statements or perhaps a three-word definition will suffice. If the goal of creative opportunity is to be achieved, then two-word function definitions are imperative. If the function cannot be defined in two words, more under-standing is required. It is a struggle to define good functions, but the result is worth the struggle.

Unfortunately, skipping over this basic effort does not create the in-depth understanding of the project required for substantial breakthroughs in technology or the creative insight necessary to develop new ideas beyond the commonplace. The importance of substantial effort in defining functions cannot be over emphasized. Poorly defined functions are analogous to building a house on sand. A weak foundation will not ensure a strong structure.

This requirement to define a function in two words is a forcing technique that requires consensus among team members, eliminates confusion, creates in-depth understanding of the requirement, clarifies overall knowledge of the project and ultimately breaks down the barriers to visualization that will lead to new, outstanding solutions to the project.

In his book, *Techniques of Value Analysis,* Mr. Miles recognized the difficulty of applying this technically simple concept. He said, "While the naming of functions may appear simple, the exact opposite is the rule. In fact naming them articulately is so difficult and requires such precision in thinking that real care must be taken to prevent the abandonment of the task before it is accomplished." He also said, "Intense concentration, even what appears to be over concentration of mental work on these functions, forms the basis for unexpected steps of advancement of value in the product or service."

Carlos Fallon[4] pointed out that just knowing what something costs can save 5%, changing materials may save 10%, but finding a better way could save 30% or more or develop a technological breakthrough and absolutely requires function analysis.

From the beginning of function analysis it was recognized that defining functions is a difficult task and that genuine effort is required to successfully define the functions of a product or service. However, if the goal of maximum creative opportunity is to be achieved, the effort must be expended. Unfortunately, there has been very little written to show how to define functions or how to identify a properly defined function.

The Link Between Function and Creativity

The theory and philosophy of function definition and analysis is simple but the application is difficult. It is so simple to understand that it is frequently mistaken for common sense. It is, therefore, necessary to constantly apply the strictest discipline to keep from reverting to the same old way. It is especially difficult to defer judgment and follow the process to completion, especially when the details begin to build up in massive amounts. In effect, the function definition takes something with which a group is thoroughly familiar and in the process of converting that something from what it is to what it does, can create utter chaos. Methods for organizing and analyzing this chaos offer maximum opportunity for creative development.

It has already been emphasized that functions must be properly defined to achieve maximum benefit. This takes time and practice; however, once the technique has been developed it becomes second nature and the opportunity to increase the effectiveness of your thinking increases appreciably. Function is not a "fill in the blanks" exercise to be completed as quickly as possible. Defining function is the foundation of VE and it requires effort to overcome perceptual bias and break down resistance to visualization to help to see beyond our usual experiences.

In the course of a typical study many functions will be developed. In the case of simple products, it is not unusual to find twenty-five to thirty functions. In a complex project, system or organization analysis, over a hundred functions may be defined. The handling and analysis of these functions through the process of function definition, cost-function analysis and value determination to find the poor value functions is complex and laborious. However, the Argus system, which will be discussed in detail in Chapter 12, helps to organize this random group of data into a rational picture of the project and in the interim, continues to create new insights into the project at the same time it is building a continuous readily recognized, interrelated system.

Define the Function

The function is the property that makes something work or sell. It is what we pay for. Function represents our needs, desires, or requirements depending on the point of view or the role being played by the analyst. The definition of function requires skill, practice, and an awareness that the function must be defined in a manner that will broaden the opportunity for creativity.

A function is not an action! It is an objective of an action. For example, "file papers" is an action. What is the objective? What are we trying to do when we file papers? The function should tell us why it is necessary to file papers, or what we want to accomplish by filing papers. It should tell us the purpose of the action which might be "store information." This then could be at least one proper definition of the function being performed. It should be obvious that there will be a number of actions that take place in performing the "store information" function. Among them will be reading, classifying, sorting, etc.

In the case of hardware, a description such as "turn shaft" is not a function; it is an action. The objective of the action must be determined as in the previous example. Why is it necessary to turn the shaft? What are we trying to do when we turn the shaft? What is the purpose we want to accomplish? The answer may be "transmit torque" or perhaps "store energy" as in a spring. The answer will depend on the application and objectives of the person or persons in the team and may well be different for the engineer, lawyer, teacher, or the role being played at the time.

It should be obvious that in both of the two previous simple examples, the function definition offers more alternatives for creative development than the action. For example, it would be possible to think of at least a dozen different ways to store information or transmit torque than to file papers or turn a crank.

So, in defining functions it must be remembered that the function is the objective of the action. It is the result to be accomplished. The action is one method to accomplish the objective.

The basic rule is to define the function in two words, a verb and a noun. This is the shortest possible statement. The resultant definition should be such that it is not restrictive in that it defines a method for performance. An abstract definition will offer an opportunity for creative questions that will produce a number of alternatives. For example, "transmit torque" not "turn screw"; or "control climate" not "cool air."

It is also important that the function be measurable in some unit term such as weight, cost, volume, time, space, etc. This is necessary in order to establish a value for the function. If we know the function costs $5.00 we can almost immediately determine if we feel it is worth the cost. In some cases, the measure may be satisfaction, desire, or some other abstract measure that will require more subjective analysis but can still be measured by comparative techniques.

Examples of some typical functions showing possible units of measure are:

Product Functions

Verb	Noun	Unit
transmit	torque	lb. ft.
conduct	current	amps
support	weight	pound
control	flow	CFS
store	energy	watts

Operations — Systems

Verb	Noun	Unit
create	design	time
confirm	design	time
authorize	program	cost
isolate	heat	cost
measure	performance	man-hours

Architecture — Construction

Verb	Noun	Unit
create	environment	humidity
control	noise	decibels
prevent	vibration	CPS
attract	attention	opinion
distribute	material	time
convert	energy	cost

It can be seen from this list of functions that the same functions may be found in different products in different industries. For example, "contain environment" might be a function of a mechanical product, a building or a system requirement to meet EPA or some other government standard. The method appears to be universal. Although the various functions may require different solutions for different industries it is always possible that the way a function is being satisfied, as in an automobile, could be modified for a building. Again, the way a manufacturing process is being performed could lead to modifications to improve an administrative system. The whole idea of function definition is to break down mental constraints and open minds to new ideas no matter where they come from.

Since function definition is somewhat difficult at first, at least until you develop the technique, I reviewed past projects to determine exactly what functions helped to develop successful new or improved recommendations. As a result, I listed examples of actual functions from hundreds of successful projects in 10 different categories covering every imaginable field of endeavor. These fields are:

1. Administration/Management Information Systems
2. Capital Equipment/Product Analysis
3. Government/Community Affairs

4. Hospitals/Healthcare
5. Manufacturing
6. Personnel
7. Product Analysis
8. Product Planning/Marketing/Advertising
9. Real Estate/Building Management/Construction
10. Technical Operations

After reviewing the function list I found that only 278 verbs and 364 nouns were used in all of the function definitions on hundreds of projects. This was a total of only 642 words. How could this be possible? Does this mean that if you learn this list of functions you would be able to solve any problem? No, it does not. However, it does mean several things; among them are these: Looking at something totally unrelated to the present project can give new insights into the problem; the problems are not different, it is the product that is different. This comprehensive list of function examples is in "Park's Catalog of Frequently Used Functions in Value Engineering" in the Appendix.

Certain key words help to break mental constraints to broader thinking. We can learn about our own projects and problems by talking to people about their totally unrelated projects. Properly defined functions can help us to see our project in a new light. These data are tabulated in the Catalog of Functions in the Appendix. However, these functions should only be used as a reference and should not be copied. As already stated several times, results come from the understanding derived in the effort.

The three rules for function definition are:

Two words — Verb, noun — The shortest possible statement
Measurable for evaluation — To offer a basis for comparison
Offer creative opportunities — The question for resolution

The reason we go to the trouble and effort to define functions is to create new insights in the way we see our problem. We frequently become so involved in the project that it is difficult to see beyond the limits we have set for ourselves or those that have been set for us. To break out of a situation we must work and struggle to develop functions that will help us see things differently. This effort frequently results in developing information that may never have been known before; this can be a very exciting and rewarding experience.

Remember: **People see what they expect to see and ignore what they don't expect.** Psychologists call this phenomena *psychological bias.*

Functions for Creativity

In defining functions it is important that several key questions be kept in mind. These are:

- What are we really trying to do when we perform this action, or what does this part really do?
- Why is it necessary to do this?
- Why is this part or action necessary? What does it really do?

Specific answers to these questions will aid in zeroing in on a useful definition. However, each question must be answered differently. Write the answers down. Each question must be answered independently; the same answer cannot be used to satisfy all three questions.

When defining functions it is also necessary to keep in mind the role being played. The functions of a product will be different depending on whether the role being played is that of the plant manager, the product, or the customer. This role playing may be difficult at first, but becomes easier with practice. The idea is to "let the job be the boss," as Kettering said. Be the crankshaft. What do you do? How do you feel. Act the part of the customer. What do you see? What does it do for you? If you were the plant manager, what would you want? How would you get it?

This system helps to eliminate bias in that functions can be identified from all viewpoints and sorted out later by using special techniques designed for that purpose. For example, recently one of my associates was attending a picnic with one of his young grandsons. During the course of the picnic the grandson challenged his grandfather to climb an apple tree to see who could climb the highest. As grandsons often do, he climbed higher than his grandfather and when he looked down he shouted, "Grandfather you're bald." True, but a revelation to the grandson who had never seen his grandfather from above before. This was a new insight; one he got from viewing the subject from a new and different angle. That's what function definition should do for you.

It must be understood that not all two-word definitions are useful as creative tools. Many definitions are what may be called industrial clichés. For example, a product or process design frequently develops a definition "heat-treat part." This is not a useful definition. For one thing, it is more than two words. Second, it is not a readily understandable definition. It is a definition that everybody understands and yet no one understands. In order to produce changes in concept, the objective of the heat treat operation must be clearly understood.

By asking one or more of the questions stated above, "What are we really trying to do when we heat treat this part?" it is possible to zero in on a useful definition. For example:

What are we trying to do?	Increase strength.
Why is it necessary?	Prevent cracking.
How can we do this?	Control cooling.

From these answers some useful, creative functions can be developed. Creative questions can now be asked. "How can cracking be prevented, or how can cooling be controlled?

From these actual responses, a new manufacturing process was created that has produced substantial product cost saving as well as lower manufacturing cost.

Another example is the term *improve communications.*

What are we really trying to do?	Achieve understanding.
How can we do this?	Exchange information.
Why?	Resolve conflict.

Again useful functions have been provided that supply a team with the material for questions that can produce a wide variety of methods to achieve their goals.

In a complex operations analysis of the controller's operation in a large company the basic function was defined as "control cost"; however, discussion indicated that this was not really true because it was possible for higher authority to overrule his advice. When the three questions were asked, the result was as follows:

What does the controller really do?	Counsel management.
How does he do this?	Advises restraint.
Why?	Improve profit.

These responses satisfied all of the members of the team and led to greater understanding of the controller's operation and increased productivity. In addition, the preconceived understanding of the operation was revised and clarified.

A word of warning; *PROVIDE* is a taboo word. It is not useful in the defining of functions that produce understanding. It is not unusual to find projects where the verb "provide" was used to define almost every function. When the function has been defined using the verb provide, in many cases it is possible to convert the noun to a verb and redefine the function. For example:

Provide space should be defined as space elements.
Provide support should be defined as support weight.
Provide transportation should be defined as transport material.
Provide control should be defined as control noise.
Provide resistance should be defined as resist wear.

Start at the Top

In defining functions first start with the assembly, complete process, program organization, or whatever the total project may be. Define the functions. At first the functions may not be the best definition possible: I call these apparent functions. In the early project stage don't haggle over whether the function has been properly defined; it can be refined later. In addition, it is difficult enough to get into the two-word habit. Write every thought down so it will not be forgotten

When all functions of the assembly have been defined, take each part or segment of the system and define the functions of each. There will be some duplication, but this will be screened out later. It is easier to screen out duplication rather than to check or discuss the subject at this time.

Types of Functions

After all functions have been defined, screen the list to eliminate duplicate functions. Now screen the list again to identify the basic function. The basic function is the function or functions upon which all other functions depend. If the basic function is not needed, none of the other dependent secondary functions will be needed.

In many cases, a number of functions beyond the system scope will be defined. These are called higher- or lower-order functions depending on whether they cause other functions or if other functions cause them. Higher-order functions cause the basic function and make it necessary. Higher- and lower-order functions are not usually a part of the project but do increase understanding of the total project by identifying the need for the basic function. The basic function is the method selected to satisfy the higher functions.

Notice that all functions used in examples are positive statements. Because of the logic involved in the overall VE system, negative functions are not acceptable. It is therefore, sometimes necessary to develop positive functions from negative questions or statements. To do this, use the Function Determination Process which involves a series of simple questions.

> Function Determination Questions
> 1. Select a problem.
> 2. Ask the problem resolution question.
>   Why is this a problem?
> 3. Ask the function determination questions.
>   What must be done to solve the problem?
>   Why must the problem be solved?
> 4. Review each question and rewrite the question in function language (2 words — verb, noun)

An example will show how the system works.

In a project involving the productivity of drafting operations one of the problems seen by management was stated as: Automatic progression deteriorates performance. This means that promotion and salary increases for draftsmen are based strictly on length of service. The answers to the problem determination questions are shown in Figure 5.1 and the list of functions derived from the list is Figure 5.2. It is interesting to note that the original problem, automatic progression deteriorates performance, has become lost. It has really turned out to be a symptom of the problem.

By application of the function definition principles cited here, the result will be clearly understandable, measurable for use in cost function analysis and function evaluation, and will lead to outstanding opportunities for creative improvement.

Some Examples of Product Functions

These examples of actual two-word function definitions used in successful projects will give you an idea of the type of function definitions that lead to outstanding creative results. They have been extracted from the Catalog of Functions in the Appendix. In each example only a representative sample of the total list developed in the actual project is shown. The examples were selected to show the diversity of ideas represented.

Now, an additional word of caution. These definitions worked for the particular team involved. The functions produced were the result of a struggle that produced insight and understanding for them and resulted in an outstanding accomplishment. Each team went through the process. It is not possible to simply copy their functions and expect to produce results for a different process. Remember, the output is directly related to the input.

1. Why is this a problem?	2. What must be done to correct this problem?	3. Why do we have to correct this problem?
No adequate award/reward system.	Eliminate automatic progression.	Increase production.
Discourages initiative.	Require higher level work for higher pay.	Improve morale.
Discourages individuality.	Establish performance standards for each level.	Reduce cost.
May be inconsistent with performance.	Develop ways to measure performance.	Improve quality.
Causes unqualified people.	Develop apprentice system.	Develop qualified people.
Increases cost.	Reduce number of levels.	Increase responsibility.
Limits merit awards.	Develop training school.	Decrease supervision work load.
Fosters mediocrity.		Provide initiative.
Doesn't recognize variations in skills.		Increase management control.
Reduces management control.		Provide executive rewards.
Lowers morale.		Establish drafting career opportunity
Senior controls promotion.		Relate rewards to contribution.
Ability is not recognized.		Establish discipline.

Figure 5.1 Automatic Progression Deteriorates Performance

List All Functions

Verb	Noun
increase	production
improve	morale
reduce	cost
improve	quality
develop	personnel
increase	responsibility
supply	initiative
increase	control
supply	rewards
indicate	opportunity
measure	contribution
establish	discipline
develop	teamwork
establish	qualification
establish	standards
reduce	work load
define	measurements
reduce	steps
establish	training
involve	principles
recognize	contribution
control	operation
improve	recruiting
define	classification
increase	incentive

Figure 5.2   Function List

Construction — Wind Tunnel

create	vibration
make	noise
control	moisture
achieve	comfort
distribute	air

Job Evaluation — Advertising Account Executive

locate	prospect
attract	attention
arouse	interest
show	benefit
obtain	commitment

Public Works — Park Lake Dredging

create	mixture
separate	solids
transport	mixture
add	energy
remove	water

Operations Analysis — Engineering Department

create	design
authorize	program
evaluate	need
confirm	design
clarify	information

Operations Analysis — Controller's Office

measure	performance
develop	plan
counsel	management
highlight	deviations
discharge	obligations

Manufacturing Process — Heat-Treating Operation

convert	energy
space	material
minimize	distortion
remove	waste
transport	energy

Summary

Although defining function may sound simple it is in fact difficult to get started, and the practice is not usually carried out. It is strange but true that it is possible to read and understand a discussion and to accept the premise but then to revert to the same old methods.

The objective of function definition is not to provide answers but to help you to ask the right questions, questions that will lead to creative opportunities that will produce outstanding results. Again, let me say that a properly defined function must satisfy the following three requirements:

- Defined in two words: a verb and a noun
- Measurable for evaluation
- Offer creative opportunities

The process does require effort, as Mr. Miles said, "Intense concentration is required." It requires the most difficult work possible — thinking — but the rewards are great. When the job has been done correctly you may be able to say, *AH HA, I SEE !* A very exciting experience.

References

1. Significant parts of this chapter are reproduced from Park, R. J., *Best Value Is The Goal* and *Social Value, A Matter of Opinion,* SAVE Proceedings, Northbrook, IL, 1986 and 1987. With permission.
2. Miles, L. D., *Techniques of Value Analysis and Engineering,* McGraw Hill, Inc. New York, 1961.
3. Webster's New International Dictionary, G & C Merriam Co., Springfield, MA, 1981, 1556.
4. Fallon C., *Value Analysis,* 2nd Ed., Triangle Press, Irving, TX, 1978, 9,

Chapter 6

Value: A Matter of Opinion[1]

Introduction

Cost is a fact; it is a measure of the amount of money, time, labor, and any other expense necessary to obtain a requirement. Value, on the other hand, is a matter of opinion of the buyer or customer as to what the product is worth, based on what it does for him. In addition, a person's measure of value is constantly changing to meet a specific situation. Unfortunately, when value is measured in dollars, confusion frequently results.

A discussion of value is like opening Pandora's Box. Once the box has been opened there are so many ways to go that the scope of the discussion must be specifically defined. Because of the breadth of the subject we will confine our remarks only to aspects of value that are directly related to cost and cost analysis. However, this will include both product value that is related to cost and social value, which frequently must be measured on a priority basis.

A discussion of value is necessary because a measure of value makes it possible to determine if we are spending our money wisely or if we are not getting good value. A measure of value gives us a measure for comparison. Although we may not understand the specific basis of value, we are, each and every one of us, constantly making value judgments. Our objective here is to show how these value judgments can be used in business and in any other situation that involves cost.

The history of value goes back at least to the Greeks. There are references that give Aristotle credit for identifying seven types of value. Other references credit Vitrovius,[2] whose writings dated before 27 AD and discuss six types of value. These six types of value are usually defined as moral, aesthetic, legal, sentimental, use, and market. I have added social value, which may be a combination of at least moral and legal and sometimes economic. I also see use and market value as part of economic value.

There are dozens of texts and references to value. In addition, the term *value* is constantly used in advertising. Many of the references include specific discussions of value, value standards, and value visibility. One, *Value — Its Measurement, Design and Management*[3] refers to value as a force that governs our behavior. It certainly does that. Another, *Value Standards,*[4] develops a set of mathematical equations to make it possible to calculate a value based on criteria such as cost, physical characteristics, quantity of a material, and the function required, to provide a value in dollars. Value standards are referred to in Chapter 12, Value Engineering.

With such a wide range of outlooks on the subject we will review some of them to obtain a workable understanding of the term. We will compare some of the elements to determine similarities and differences and set up a basis for practical application.

Definitions

Webster's New International Dictionary[5] offers 12 definitions for value. The four selected here apply to our situation. Value defined as "1. A fair return in goods, money, or services etc., for something exchanged, 2. Monetary worth of a thing, estimated or assessed worth of a thing, 4. Estimated or assessed worth; valuation, 10. Economics (b) Proper price; the quantity of money, goods, or services, which an article is likely to command in the long run, as distinct from its price in an individual instance. (c) The estimate which an individual places on some of his possessions as compared with others, independently of any intent to sell."

So Webster says value may be measured in dollars or just by comparison with something else. A new word is also introduced, *worth*; monetary worth. The *American Heritage Dictionary*[6] defines value in much the same words but adds: monetary or material worth, worth in usefulness or importance to the possessor; utility or merit. Figure 6.1 illustrates the definition of value.

The new term worth is defined as the quality of something that renders it desirable, useful, or valuable; the material or market value of something;

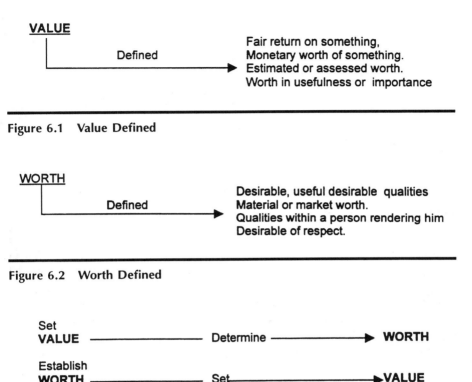

Figure 6.1 Value Defined

Figure 6.2 Worth Defined

Figure 6.3 Value–Worth Relationship

the quality within a person that renders him deserving of respect. Worth is defined in Figure 6.2.

So value is defined in terms of worth and worth in terms of value. In other words, if we set a value on something we determine its worth. If we establish the worth of something, we declare its value. For all practical purposes, the two words are interchangeable. This relationship is illustrated in Figure 6.3.

However, the *American Heritage Dictionary* qualifies the exchange by noting that they are largely interchangeable when the reference is monetary. Otherwise, worth is especially appropriate in denoting qualities in persons or things that add up to moral excellence or to merit, considered as an interchangeable apart from utility. Value is most often applied to what is demonstratively useful. Figure 6.4 identifies this exception to the interchangeability of the terms value and worth.

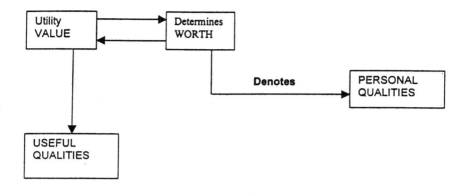

Figure 6.4

Carlos Fallon[7] states that worth is the simple concept. It becomes value when it is related to cost. He further states that cost is a necessary component of value. This tends to agree with the *American Heritage* definition. However, using both "value" and "worth" has tended to create confusion whenever the two terms are used. Therefore, in this text the term *value* will be used without reference to *worth* when at all possible.

Chris O'Brien[8] says that value is the ratio of worth to cost. Since worth is defined as an appraisal of the properties of a product it is essentially an appraisal of the function. In other words, value is the ratio of function to cost. In this case function is defined as what the product does for the customer, and could also be considered to be product performance in many instances.

Although value is frequently defined in terms of dollars, value is not cost. As explained in Chapter 4 on cost, cost is a fact that is directly related to the product or service. Value is an opinion and is related to a want or a need — what the product or service does for you or its performance to expectations. It is also important to recognize that value is subject to rapid and measurable changes as conditions change.

Since value is related to function, the first step in determining value is to determine what the product or service does, its function. What is its purpose? A value cannot be set for a product; it can only be set for what the product does for the customer, its function.

If we set values on a function, the function may be measured in some practical unit or it may be a relative value established by comparison with other functions. The term value can then be used as a reference for products, services, or abstract requirements without confusion.

With this understanding we can determine the qualities we would like to have in an organization, product, service, or person. A value for each can be determined by means of a ranking scale which may be in dollars, time, power, or just by determining the weight of one vs. another. If we feel that something is of value to us, we can determine its relative value by comparing it to something else or to a group of things.

Measuring Value

To review the literature regarding value we went to the Proceedings of the Society of American Value Engineers, an organization dedicated to disseminating information on value and teaching methods to obtain good value. Some of the most recent papers on the subject were by John Bryant,[9] Ron Harris,[10] Jerry Kaufman,[11] Marvin Wasserman,[12] and Carlos Fallon.[13] Each of these people has conveniently defined value as a relationship or a ratio. I have listed them for convenience.

Bryant   *Value* $V = \dfrac{Wants + needs}{Resources} = \dfrac{Selffunctions + Usefunctions}{Dollars + People}$

Harris $V = \dfrac{Worth}{Effort}$

Kaufman $V = \dfrac{Function}{Cost}$

Wasserman $V = \dfrac{Function}{Cost} = \dfrac{Utility}{Cost} = \dfrac{Performance}{Cost}$

Fallon $V = \dfrac{Objectives}{Cost}$

In each case, the denominator is a unit that can be measured in dollars, effort, people, resources, etc. So we can say that each agrees that the denominator is cost. However, there does seem to be some variation in the numerator: wants, needs, worth, function, objectives. In the discussion on function it is pointed out that function is a goal, a need, a want, an objective. That leaves worth. Carlos Fallon and Chris O'Brien describe worth as an appraisal of the properties that make a product useful, or the monetary unit of utility. In other words, the monetary unit of function. It can be said, therefore, that they all really agree that the basis for value is the ratio of function over cost.

$$Value = \frac{Function}{Cost}$$

In other words, a measure of value is how much we may have to pay for something we need or want or perhaps, as Bryant and Wasserman say, a combination, in some degree, of both. The higher the cost, the lower the value unless the function improves. However, for companies, it is dangerous to assume that increasing the function of a product or service may make it possible to increase the value, especially at a higher cost.

In 1979 the American Can Company accepted the premise that quality sells, products can always be improved, and the customers will pay more if they perceive good value. So they designed a strong paper towel called BOLT. It looked and performed like cloth and was a total failure. The consumer did not perceive a benefit or an increase in value in a paper towel that could be washed rather than discarded.

However, if we do not have cost to make a comparison, how can we establish values? Cost then must be replaced with something else which we will call a measure of comparison or opinion.

We now have two equations for value. The first may be compared on a monetary basis, and the second must be compared on a relative basis using opinion. The equation now becomes:

$$V = \frac{Function}{Cost} = \frac{Function}{Opinion}$$

Function in Value Engineering/Analysis

Mr. Lawrence D. Miles[14] developed the concept of Function which he defined as a want or a need, something we are willing to pay for. Function is the foundation on which the Value Engineering/Analysis method is built and was discussed in Chapter 5, Function: The Foundation of Clarity. Its use and application will be covered in Chapter 12, Value Engineering: A Total System. Functions may include a mix of needs and wants as noted by Bryant and Wasserman. Understanding that there are both wants and needs and that they are different is of utmost importance to manufacturers and sellers.

Recognizing that there are different ratios of wants and needs in every person has resulted in a wide choice of products and services and has made many people wealthy beyond their wildest dreams.

During the 1930s, there was a slogan by a major Trans-Atlantic steam-ship company that said, "Getting there is half the fun." It was for some people but not for others. During the days of the great migration of people to the U.S. from Europe, great steamships moved thousands of people from one place to another in great comfort or in miserable discomfort, depending upon whether you were traveling in first class or steerage. All of the passengers were moving in the same vehicle and at the same speed from the same place to the same destination at greatly varying degrees of comfort.

Those who could travel first class felt that it was worth what they were paying. Those in steerage felt it was worth putting up with discomfort for the price they were paying to achieve their goal. In both cases, it was what each could afford. We have all heard about the great comforts and convenience of the liners, but it was the thousands of steerage passengers who made the lines great.[15]

Then along came a German ship owner named Ballin.[16] He recognized the poor conditions under which many people sailed and reasoned that if he was to improve the living conditions of steerage passengers they would tell their friends and they would sail on his ships rather than his competition. This was true and did happen; however as a result the other lines improved their conditions, thereby raising the value standard.

From this it can readily be seen that it is possible to pay substantially more money for essentially the same product if we only look at the transportation. However, people pay more if it satisfies other needs, wants, desires, and they have the money.

Interestingly a similar condition exists in today's airline travel. People do not necessarily ride first class because the trip is usually short and "it is not worth the difference." So first class is usually not considered a good value unless the time is considered. Even then most people feel they can put up with some discomfort for 8 to 10 hours.

Value Classifications

For practical purposes when working with value it is beneficial to break the term into several classifications. Best value is defined as:

Value — The lowest cost to provide a function.

This value unit is then broken down into several categories.

Use Value — Properties that make something work or sell.
Esteem Value — Properties that make something desirable to own.
Exchange value — Properties that make it possible to exchange one thing for another.

These elements of value refer to examples listed in literature published by the Society of American Value Engineers in Proceedings and other literature published by the society.[17]

In the transportation example above, the value relationships are: The lowest cost for the ticket from Europe to the U.S. is the best value: This is the use value, the transportation part of the ticket. The esteem value is the difference between the amenities over transportation for first class. The exchange value may vary depending on current economic conditions. If for some reason you could not make the trip, you might want to sell the ticket to avoid a loss. If all the berths were taken you might be able to sell the ticket for more than you paid for it. If space was available you might have to offer a discount.

Several years ago I ran into an interesting example of exchange value. I was invited to attend the annual meeting of a large realty company for whom I had done considerable work. This was at a time when economic conditions were rapidly declining and the company had decided to sell off the "land bank," property that had been purchased for future use. In most cases this property was to be used for future automobile dealerships.

Several executives who had disposed of the most property in the shortest time were formed into a panel to explain how they had accomplished their feat. After the presentation, the corporate treasurer asked why they had to sell all this property at a loss? Couldn't they have sold it at a profit? The person who had disposed of the most property responded, "When I bought the property no one said I might have to sell it. If they had, I would not have bought it."

So let us look at the situation from a value standpoint. Of the total cost of the property, the amount allocated to the necessary size of the plot would be the use value. The extra cost paid because of a special location would be the esteem value. However, there was no intent to sell the property at the time of purchase so the future exchange value was not considered. At the time of sale the exchange value was considerably less than the purchase price because of market conditions and deterioration of the neighborhoods.

The purchase of a wrist watch is another example that may be more clearly understood. The first step is to determine the function we want; let us say it is "tell time." What is the best value or the lowest cost we know of to reliably provide the function "tell time"? There are several $10.00 watches that will reliably tell time. So, our value for the tell time function is $10.00.

Why then do people pay $100.00, $200.00, even several thousand dollars for a watch to "tell time"? One reason might be because you are a scuba diver and you need a waterproof watch. Another reason might be that you are a navigator and you are looking for greater reliability and accuracy. It might also be, as Patek Phillippe says in their ads, that your watch tells you something about yourself. You can afford it, and it tells others you can afford it.

If we look at this situation, we see that the use value is $10.00 but the esteem value may be several thousand dollars. I once saw a watch with only an hour hand; its price was $6,000.00. Who would pay $6,000.00 for a watch like this? Someone who wanted to tell you that he didn't have to be anywhere exactly on time. When he got there the meeting started. Ridiculous? Maybe, but we must understand what motivated the customer. A good value to one may not be a good value to another.

A person may also buy an expensive gift for another person, for example, a $2,000.00 watch as an anniversary present. Again the cost over $10.00 becomes esteem or sentimental value. Let us assume the gift doesn't run. The receiver will probably wear it anyway because the "tell time" function becomes secondary to "create impression" or perhaps "express sentiment."

Value Is a Motivator

If we determine that something has value, we should be able to determine what makes it valuable. It may provide a service, create an impression, improve operations, satisfy a want or a need, or some other motivator.

These motivators may be degrees of satisfaction. How fast or comfortably do you want to move from one place to another? Do you want to move alone or is it acceptable to move in a group? How much inconvenience will you tolerate, etc.?

In many cases, to determine value and its component degrees, we have to understand the basic principles of motivation. To begin with, motivation comes from within a person. A person may move to obtain something he wants or to avoid something he doesn't want.

Understanding these motivations is a complex and difficult process which is subject to great risk. For example, the history of the automobile business has shown that although the economic forecasters have been able to determine the number of cars expected to be sold with reasonable accuracy, no one has been able to determine whose cars or what models will sell and which ones will not. As a result, the industry runs into periodic catastrophes. They do not know how to determine the customer's value system. This is somewhat understandable as it may change from day to day or even from minute to minute. Another part of the problem is that the designers and manufacturers think the customer has the same value system they do. Frequently they do not.

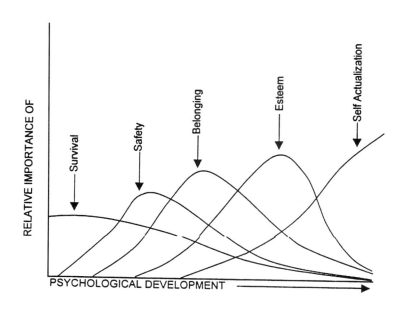

Figure 6.5   Maslow's Hierarchy of Needs. (Reprinted from Maslow, A., Maslow's Hierarchy of Needs from *Individual in Society*, Kresch, D., and Crutchfield, R. S., McGraw Hill Inc., New York, 1962. With Permission.)

Consumer products, industrial products, administrative programs, and organizations all are affected by the same motivational systems that affect people. Have you ever noticed that people, especially companies, may not have enough money for the things they need but they always have money for the things they want? This is a measure of their value systems often called priorities.

Abraham Maslow[18] developed a simple scale to define the psychological needs of people. He called this scale the "Hierarchy of Needs," (Figure 6.5). This scale and Maslow's theories have been refined and developed in more detail by Dr. Clare Graves.[19]

Dr. Maslow said that people are motivated to do different things at different levels of psychological development or different levels of society. He divided these motivational factors into five basic needs. As each need is satisfied, other higher needs arise. Although the lower level needs may never disappear, they become weaker or less important. In fact, a person has several needs at the same time but one is dominant.

Dr. Graves suggests that when any type of communication is directed toward an individual his response is essentially preconditioned by his value system. An interesting example, related to the Maslow hierarchy, shows how experience indicates how people operating on the various levels might be expected to react.

Level 1. PSYCHOLOGICAL SATISFACTION (SURVIVAL)
"Why ask me, I don't matter."
"What I think doesn't matter."

Level 2. SAFETY
"What I can work for, own and get is right."
"If others want something, let them work and fight for it."

Level 3. BELONGING
"What is good for the group is good for everyone."
"I don't like it but the group wants it so I'll go along."

Level 4. ESTEEM
"What I want is right."

Level 5. SELF ACTUALIZATION
"What the facts and evidence show is fair."

Immediately following the settlement of a strike at a large company in our area, the press was interviewing a striker as to how he felt about the new contract. His reply was, "I don't really like it but that's the way the members voted so I'll go along with it." Clearly he fell into the Level 3 Belonging category.

Sometimes substantial changes can occur in an individual's psychological development. In other cases, individuals may become emotionally blocked and become "frozen" in the same system for a lifetime.

It is important to note that an individual may operate on more than one level at a time — for example, on the job, at home with the family, or in some hobby or specialized avocation.

Although Maslow's Hierarchy readily explains basic motives, the reasons that people take action are based on much more complex conditions. Among them, professor Lloyd Warner's[20] theories of social classes are important factors in developing actions (Figure 6.6).

These values change, even among people of great wealth. In a recent *Fortune* magazine article,[21] the changing values of the Rockefeller family are revising the structure of the fortune. The fourth and fifth generations have totally different values than those before them. In many cases, they want more money to spend.

CLASS AND CULTURE (WARNER)

UPPER - UPPER	OLD WEALTH
LOWER - UPPER	NEW WEALTH
UPPER - MIDDLE	MGRS. PROF. — SALARY, FEES
LOWER - MIDDLE	SUPERVISORS
UPPER - LOWER	WAGES
LOWER - LOWER	UNEDUCATED

Figure 6.6   Class and Culture

The basic theory is that people take action based on the desire to obtain something they want or to avoid criticism or punishment. They may not understand all of the factors that cause them to be motivated in one direction or another, but a basic knowledge can cause you to recognize why each person may react differently to a situation and why he has different values or priorities.

These values or priorities are developed based on long-term experience. It is often difficult for a person with one set of values to understand why a person with another set of values may do something. For example, in the purchase of a piece of equipment, a manufacturing person may be interested in production; an engineer in quality; a purchasing person in price; a maintenance person in reliability. These priorities are set based on their goals. The action they take is based on each person's set of values. However, the purchase should be made based on what is best for the operation. The affected operation representatives should be brought together to develop a consensus as shown under how to measure values.

During 1965 and 1966 a serious investigation into the impact of current technological advances on American values was made under a grant from the Carnegie Corporation and IBM Corporation. The investigation was centered at the Department of Philosophy at the University of Pittsburgh.[22] The intention was to find ways to guide social change in directions compatible with the realization of our deepest values.

They developed an interesting definition of value. It was defined as follows:

> A value is an attitude for or against an event or phenomenon, based on a belief that it benefits or penalizes some individual, group, or institution.

Although the group was made up of 17 eminent persons from a number of major universities and industry they immediately ran into two major obstacles. The first was the need to draw up an inventory of the values of an individual or a group; the other, the determination of the soundness of the values to which individuals or groups subscribe.

Models were constructed and developed, but the group found it impossible to look at the problems without applying their own values. They were not able to put themselves in the position of underprivileged people. The consensus was, however, that it may become possible to predict the effect of change on values and even look forward to professional value forecasters in the future.

Define the Value

I said in the introduction that a discussion of value becomes extremely complex, and the more the subject is developed the more complex it appears to become. However, the more involved we become in a wider scope of VE projects the more likely it is that we will need to understand the basic philosophy of value.

We set values without much thought every day on products and services as well as things that concern us. However, in the case of a specific project we concentrate on the value of the functions to be performed because these are what we want or need and what we pay for. We try to determine what the function is worth to us.

The best way to determine the value of a function is by comparison to another function we know is a good value. If we are working with specific commodities over and over, such as steel tubes, castings, or perhaps forces, we might establish a set of value standards or value indexes based on a series of tables or charts. However, since costs are constantly changing, the standards will also change and it will be necessary to keep them up to date. There is a procedure to be followed to develop individual values, and they can also be used to develop standards as well. The procedure requires an understanding of the term *function* and how to identify and define functions, which was discussed in Chapter 5.

The first step in determining value is to define the function that is being performed or that we want. The method described here can be used in evaluating the functions of products or services. However, there are cases where other methods may be required, as is discussed later in the discussion of social value. As an example we will set the value for a simple component used to conduct electricity in electrical receptacles, switches, etc.

In the example in Figure 6.7 the conduit (wire) is connected to the conductor by the screw. Current is conducted through the conductor and the rivet to the contact. Cost for each piece is shown. The basic function

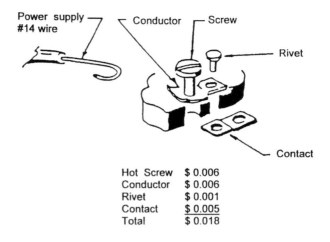

Hot Screw	$ 0.006
Conductor	$ 0.006
Rivet	$ 0.001
Contact	$ 0.005
Total	$ 0.018

Figure 6.7 Power Supply Contact

of the system is to "conduct current." Although the screw does not "conduct current" it is part of the system. The screw actually "maintains contact" between the power supply and the conductor. The system costs $0.018. The function is "conduct current." What is the value of the system, the lowest cost to reliably provide the function "conduct current"?

What else will do the job and what does that cost? — One of the twenty basic techniques.

The problem is to conduct current about one inch. It is not necessary to conduct more current than the wire can supply. For this system to conduct 110 volts, number 14 wire is adequate. Number 14 wire costs about $0.007 per foot or $0.00058 per inch. The wire will reliably conduct the required current one inch for $0.00058; this then is the value of the system. It is only 3% of the original cost. This value does not consider "maintain contact," which may be eliminated. It only considers "conduct current" and is a perfectly good value for the function. The value $0.00058 covers only "conduct current"; other functions such as "maintain contact" or "break flow" will be considered individually. By moving step-by-step through the functions of the system and setting a value on each individual function, the value of the total is gradually developed.

It is important to recognize that the value is not necessarily the goal. It is a means to determine the imbalance in the system, the areas that do not provide good value are determined by major differences between function cost and function value. These functions, when identified, precisely determine the places to concentrate creative effort for maximum benefit. They highlight the targets for opportunity in the system.

The sum of the individual functions determines the function value of the total system or what the system should cost — the target value or goal for achievement.

Social Value — A Matter of Opinion

All problems and projects are not clearly defined, nor are they specific. In fact many are often described as a fuzzy mess, and they must be clarified and defined, at least to some degree, before the value can be defined. The more nebulous the problem the more likely that the Argus system, a method for charting functions based on the Theory of Function relationships, can help to define loose thoughts and organize them so they can be evaluated and problem areas precisely defined. Details of the system are discussed in Chapter 12.

Social problems are often not clearly defined. More often than not they fall into the fuzzy mess category. For example, administrative and organizational projects such as increasing the productivity of emerging nations, aiding in the improvement of race relations, juvenile delinquency, social services, including various phases of welfare programs, and employee and job evaluation are a few hazy area projects that were not clearly defined. However, they all responded readily to the Argus system approach which is explained in detail in Chapter 12.

In a typical problem with well over a hundred functions, how can we determine where to concentrate our creative effort for maximum benefit? In a product or service we can use economic measures such as dollars, man-hours, or other measurable criteria. How can we set a measure on ethics and morals? All through history great minds have been discussing and struggling with the problems of social values involving ethics and morals to set standards for social performance. What can we do to improve the situation?

Value in the Past

James Boswell[23] reported in his journal on a discussion with the great Dr. Samuel Johnson and others in July 1763. Johnson quoted James Petty, a 17th-century political economist, as saying, "If a man has three pounds per year it is enough to fill his belly, shelter you from the weather, and get you a good strong leather coat of bull's hide. Beyond this is artificial taste." As we would say, esteem value is the measure of all wants after the needs have been satisfied.

VALUES

Top Five	Bottom Five
Self respect	Pleasure
Family security	World of beauty
Freedom	Salvation
Accomplishment	Social recognition
Happiness	Equality

ATTRIBUTES

Top Five	Bottom Five
Honesty	Obedience
Capability	Cheerfulness
Ambition	Politeness
Independence	Helpfulness

**Figure 6.8   Values and Attributes:
How Managers Rank Personal Values**

Values and the Future

As noted earlier, in 1965 and 1966 a serious investigation into the impact of technological advances on American values was conducted at the University of Pittsburgh to find ways to guide social change. As already pointed out, the group was not able to develop a successful conclusion to the study because no one in the group was able to set aside their value system to put themselves in the position of people less privileged than themselves, none of which were members of the group.

There is substantial disagreement on whether it is human nature or corporate culture that affects social values. However, studies are under way at the University of Pittsburgh, Washington State University, Santa Clara University, and other groups and organizations, as the result has major effects on business.

The chart in Figure 6.8 lists values frequently included and excluded from company codes of ethics. It is interesting to compare the personal values of Figure 6.8 with the corporate values in Figure 6.9. I would suggest that you use the methods illustrated in this text to determine how your sense of values compares.

Included	Frequency
Relations with US Government	86.6%
Customer/Supplier Relations	86.1
Political Contributions	84.7
Conflicts of Interest	75.3
Honest Books or Records	75.3

Not included	Frequency
Personal Character Matters	93.6%
Product Safety	91.0
Environmental affairs	87.1
Product Quality	78.7
Civic and Community Affairs	75.2

Figure 6.9   Company Codes of Conduct

Social Problems and the Argus System

A number of years ago we were looking for ways to learn the potential power and scope of the Argus System. In our search we found several community problems that were crying for improvement. Working on these problems offered an opportunity to develop the Argus System and at the same time educate a number of people in the Detroit Value Engineering Society in our methods.

The first was a two-project study for the Royal Oak, Michigan, school board. They had lost a series of tax millage elections and needed to improve school–community relations. A second problem was to relieve congestion in the high school during class change periods. The result of this project is shown as an example in Chapter 14, Examples: Winning Results.

A program was set up with a group of Value Society members as leaders and selected citizens as team members. The results were outstanding. The team lessons learned were then brought into our respective companies and applied to projects to improve productivity, define building concepts, design new organizations, administrative systems, sales strategies and business plans. Many of these projects were reported in Society of American Value Engineers Proceedings.[24]

As time progressed we conducted more and more complex projects and improved our methods until we felt we had developed a practical package that has been proven extremely effective in solving product problems as well as social problems that occur in every organization.

Sometime later a management development specialist in a major automotive company sent a man to see me to discuss a problem and a potential project. The man was a candidate for a Master of Business degree and had been assigned a project by his counselor. It was suggested that he discuss it with me. The project was to make a cost effectiveness study of a major county youth assistance program. I suggested that he use our methods to evaluate the program. Although the man had never heard of our systems we coached him through the program which was very complex and involved. For his effort he received several awards, and we received further proof that our system worked. The diagrams and models developed during the study have been used to develop business plans and operating budgets for several years.

From knowledge such as this we expanded our operations in one of the big three automotive companies to the solution of OSHA problems, productivity improvement in administration and manufacturing operations, material procurement and distribution, the identification of the characteristics of high-impact, low-cost product changes, why people buy cars, future service plans, the identification of dealership building concepts, and a myriad of other projects that yielded substantial improvement.

Measuring Social Values

In technical projects involving products or manufacturing operations it is usually possible to determine piece cost or time for each operation. This makes it possible for us to set a measurable cost on each function. The greatest difference between function cost and function value is the target for opportunity[8] (TFO). These differences between function cost and function value (TFO) offer the maximum potential benefit from applied creativity. We usually do not have this luxury in determining social values. However, functions can be ranked and weighted to achieve a hierarchy of values.

Previously we have shown that value is defined as the ratio of function to cost. However, if we do not have cost to help us make a comparison we must use something else, which was called a measure of comparison or opinion. Value was thus defined as:

$$V = \frac{Function}{Cost} = \frac{Function}{Opinion}$$

This means that the useful qualities of one function can be compared to the useful qualities of another function and judged by an opinion of the personal importance of one vs. the other. A person can then make a judgment between two functions as to which is more important to him. This decision will be based on his background and experience. This tends to agree with the definition defined at the University of Pittsburgh in 1975.

How can a group of people with diversified backgrounds rank a hundred or more functions much less weigh them? They can do this by using some relatively simple processes, Pareto Voting[25] and Paired Comparisons.[26] By using these methods sequentially, a ranking and weighting can be developed for the functions of a complex system by a group of people with diversified backgrounds and widely varying education. The first step in the process is Pareto Voting. In Pareto Voting each member of the group individually selects 20% of the functions that he feels are the most important or that have the highest impact on the subject in question. These are then consolidated into a team list. If the total number of functions to be ranked is 100, each person would select 20. It might seem that as a result a team of 5 would select a total of 100 functions. This is not the case. Although the final list would contain over 20 functions and could contain over 100, many of the functions receive two or more votes, thereby making it possible to quickly identify those that appear to be the critical functions. Based on the number of votes per function, a decision is made to drop all functions with less than three votes. These remaining functions will be the most important. It is best to retain 7 to 10 functions on this critical list.

It is usually very difficult for people to organize a list of even 10 items in order of importance. However, it is possible to make a decision between two items. If the choice is to select one of two it may be difficult, but it can be done. Once the decision is made, the selected choice immediately achieves a higher value than the other. This is the basic principle behind the simple ranking and weighting system called Paired Comparisons and is the second step in the rating process.

Paired Comparisons compares a list of items to rank and weigh them. Ranking is the assignment of a preferred order of importance to a list of items. Weighting is the determination of the relative degree of difference between the items.

In Paired Comparisons each item is compared in turn to every other item on the list using a simple matrix. A comparative decision is made between two items on a two-level basis; there is either a great difference or a minor difference. The decision can be made on the length of time it takes to decide. If there is no question as to which item to select, there

Key letter	Alternatives	Weight
A	Majorca	
B	Florida	
C	Colorado	
D	Caribbean Cruise	

Figure 6.10   Table of Alternatives

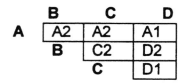

Figure 6.11   Evaluation Grid

is a great difference. If the decision is difficult or requires thought, the difference is minor. A major decision is rated a 2; a minor decision is rated a 1. A simple example will illustrate the process.

It is desired to select a vacation from these areas: Majorca, Florida, Colorado, or a Caribbean Cruise. The choices are shown in Figure 6.10.

Start by comparing A, Majorca to B, Florida. In the example the group immediately chose Majorca. The difference was, therefore, great and the choice was rated a 2. This is then placed in the proper box in the Evaluation Grid (Figure 6.11). Next, compare A to C and A to D. The responses are shown in Figure 6.12. The comparison of A to D was not an immediate decision. It required some thought and was therefore rated an A1. Now, drop the A and compare B to C, Florida to Colorado. The group chose Colorado and rated it a C2. After comparing B to C and D, drop the B and compare C to D. Notice the widely different choices as each step is taken in moving through the matrix. After completing the comparisons the characters in the grid are totaled and the result is shown in Figure 6.12.

By following the two simple processes of Pareto Voting and Paired Comparisons the group studying the juvenile delinquency problem developed the critical list of functions shown in Figure 6.13.

Rank	Alternatives	Weight
1	Majorca	5
2	Caribbean Cruise	3
3	Colorado	2
4	Florida	0

Figure 6.12 Evaluation Summary

Function		Weight
Verb	*Noun*	
Protect	Children	239
Resolve	Conflict	219
Enhance	Growth	193
Effect	Change	183
Generate	Trust	167
Accept	Problems	155
Enhance	Life	153
Understand	Problems	149
Involve	Community	149
Adjust	Attitude	143

Figure 6.13 Critical Functions — Study of Juvenile Delinquency

A detailed example showing how to set values on a system of functions that had been identified in a study is illustrated in Chapter 14, Examples and Illustrations.

By using the same methods, Figure 6.14 shows a list of the 10 most critical functions developed by a team studying the design of future automobile dealerships. The functions shown in Figure 6.14 were given to three architects as a competition. The architects were asked to consider the function list in preparing their designs and to design a facility to satisfy the required functions. The architect that developed the most innovative solution to the largest number of functions won the competition and was awarded the contract for the design.

Function		
Verb	Noun	Weight
Attract	Prospect	201
Sell	Vehicles	180
Retain	Customers	180
Manage	Operations	174
Sell	Service	153
Exchange	Goods	151
Train	Personnel	149
Sell	Components	142
Design	Facility	127

**Figure 6.14   Primary Functions —
Automobile Dealership**

Summary

The need for a way to clarify and solve complex, poorly defined problems is great. These can run from neighborhood problems to major individual crises as well as problems in industry, in manufacturing, and administration. The need to know the important things in a situation is paramount. In other words, it is important to set values on wants and needs. This requires an understanding of value, how to set values and rank them to determine the things that are important from lesser value elements.

This is not always easy to do. In most cases we have found that people find it very difficult to rank and weight a series of items. The examples in the text should help. Just in case you don't believe this is difficult, a simple problem developed by NASA is included in Figure 6.15 for you to try. This problem is included here for you to try to see how well you can do in organizing a group of random data. You may already have a system you can use. If you do not, try the puzzle, then try the techniques described here.

The NASA solution to the Lost on the Moon Problem is shown in Figure 6.16. Your score is the sum of the differences between your choice and the NASA score for each item, ignoring whether the sign is positive or negative.

LOST ON THE MOON

Your spaceship has just crash landed on the moon. You were scheduled to rendezvous with a   mothership 200 miles away on the lighted surface of the moon, but the rough landing ruined your ship and destroyed all the equipment on board, except for the 15 items listed below.

Your crew's survival depends on reaching the mothership, so you must choose the most critical items available for the 200 mile trip. Your task is to rank the items in terms of their importance for survival. Place number one by the most important item, number two by the second most important, and so on through number 15, the least important.

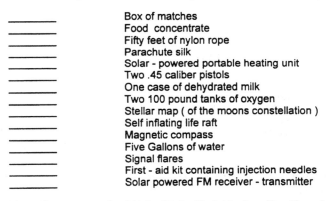

	Box of matches
	Food concentrate
	Fifty feet of nylon rope
	Parachute silk
	Solar - powered portable heating unit
	Two .45 caliber pistols
	One case of dehydrated milk
	Two 100 pound tanks of oxygen
	Stellar map (of the moons constellation)
	Self inflating life raft
	Magnetic compass
	Five Gallons of water
	Signal flares
	First - aid kit containing injection needles
	Solar powered FM receiver - transmitter

You and four other persons should take this test individually, without knowing each other's answers, then take the test as a group. Share your individual answers and reach a consensus - one ranking for each of the 15 items that best satisfies all group members. You should be aware of group decision making practices prior to taking the test as a group.

NASA experts have determined the best solution to the task.

Figure 6.15   Lost on the Moon

NASA PROBLEM SOLUTION

15	Box of matches
4	Food concentrate
6	Fifty feet of nylon rope
8	Parachute silk
13	Solar - powered portable heating unit
11	Two .45 caliber pistols
12	One case of dehydrated milk
1	Two 100 pound tanks of oxygen
3	Stellar map (of the moons constellation)
9	Self inflating life raft
14	Magnetic compass
2	Five Gallons of water
10	Signal flares
7	First - aid kit containing injection needles
5	Solar powered FM receiver – transmitter

Scoring for individuals

0 -25	Excellent	46 - 55	Fair
26 - 32	Good	56 - 70	Poor
33 - 45	Average	71 - 112	Very poor. Suggest possible faking or use of earth bound logic.

Figure 6.16   NASA Problem Solution

References

1. Significant parts of this chapter are reproduced from, Park, R. J., *Best Value Is The Goal* and *Social Value, A Matter Of Opinion,* SAVE Proceedings, Northbrook, IL, 1986 and 1987. With permission.
2. Woods, B. M., and De Garmo, E. P., *Engineering Economy,* The Macmillan Company, New York, NY, 1942.
3. Shillito, M. L., and De Marle, D. J., *Value, Its Measurement, Design & Management,* John Wiley & Sons, Inc., New York, NY, 1992.
4. Fountain, R. E., Value Standards, in *SAVE Proceedings*, Kaplan, M., Ed., 1963, 60.
5. *Webster's New International Dictionary,* G & C Merriam Co., Springfield, MA, 1981, 2530.
6. *American Heritage Dictionary,* American Heritage Publishing Co., Inc., and Houghton Mifflin Company, Morris, W., New York, NY, 1969, 1414.
7. Fallon, C., *Value Analysis,* 2nd ed., Triangle Press, Irving, TX, 1978, 25.
8. O'Brien, B. C., Unpublished management letter, *Perspective,* Oct. 1986, Park, R. J., Ed.
9. Bryant, J., Customer oriented value engineering, *Value World,* Vol. 9, No. 1, 7.

10. Harris, R. L., A guide to value consciousness, *SAVE International Conference,* Unpublished Paper, 1968.
11. Kaufman, J. J., *Executive Overview,* Cooper Industries, Houston, TX, 1981.
12. Wasserman, M., Sales market value, in *SAVE Proceedings,* Park, R. J., Ed., Southfield, MI, 1977, 105.
13. Fallon, C., *Value Analysis,* 2nd Ed., Triangle Press, Irving, TX, 1978, 24.
14. Miles, L.D., *Techniques of Value Analysis and Engineering,* McGraw Hill, Inc., New York, 1961.
15. Maxtone-Graham, J., *The Only Way to Cross,* 1st ed., Collier Books, New York, NY, 1978.
16. Ballin, A., *The Only Way to Cross,* 1st Ed., Collier Books, New York, NY, 1978.
17. Standard Definitions, published in *SAVE Proceedings* and other SAVE publications, Save International, Northbrook, IL, 1963 to 1998.
18. Maslow, A. Maslow's Hierarchy of Needs, from *Individual in Society,* Kresch, D., and Crutchfield, R. S., Eds. McGraw Hill Inc., New York, 1962. With permission.
19. Graves. C., *The Psychological Map,* National Values Center, Denton, TX.
20. Adapted from Warner, W. L., Six class system, *Consumer Behavior,* Bennett, P. D., and Kassarjian, H. H., Eds., Prentice-Hall, Inc., Englewood Cliffs, NJ, 1972, 113.
21. Loomis, C. J., The Rockefellers, end of a dynasty, *Fortune,* Aug. 4, 1986, 26.
22. Bair K., and Rescher, N., *Values and the Future,* The Free Press, Division Of The Macmillan Company, New York, 1969.
23. Pattle, F. A., *Boswell's London Journal 1762 to 1763,* McGraw Hill, Inc., New York, 1950, 314.
24. *SAVE Proceedings,* SAVE International, Northbrook, IL, 1973, 1974.
25. Shillito, M. L., Pareto voting, in *SAVE Proceedings,* Fowler, T. C., Ed., Smyrna, GA, 1973, 1312.
26. Mudge, A. E., Direct magnitude estimation, in *SAVE Proceedings,* Vol.2, 1967, 111.

Chapter 7

Quality: A Major Component of Value[1]

Introduction

There has been little doubt that quality has been the watchword in recent years. The names of Deming, Juran, Crosby, and the Baldridge Award, TQM, and other acronyms have made quality almost a household word. Now we are hearing that the 21st century will be the Century of Quality. If this is to be true, we must begin to realize that quality for quality's sake is not economical or sensible. However, like many management systems that have become popular in recent years, there seems to have been little or no understanding that quality does not stand alone; it is part of an overall equation.

An article in *Business Week*[2] entitled "Quality — How to Make It Pay," reinforces this point by noting that a number of our largest and most influential companies have found that their quality operations have not only created substantial expense but in some cases, have been found to be very unprofitable. Emphasis on improving quality does cost money, but it should result in reduced cost in other areas as well. Improved quality should produce a number of tangible benefits such as reduced scrap, a lower number of rejects, less rework, and reduced warranty claims and the resultant labor. In addition, a number of intangible factors should also result; among them would be improved customer satisfaction. However, excessive quality does not benefit anyone, most of all the manufacturer.

Every salesman knows that people do not buy a product, they buy satisfaction. They buy the performance or function of the product or service: what it does for them. They also consider the cost.

If two or more similar products are available on the market, the customer will consider what the product does, the performance, and its price. In effect, what they are doing is considering the cost relative to the function, or in other words evaluating the product based on how well it satisfies their wants and needs. The customer wants the best value for his money; satisfaction at a reasonable price. The best value is satisfactory performance for the least amount of money. We should all recognize this relationship if improved quality is to be beneficial to the entire product chain.

Note that this implies that quality is in the eyes of the customer. Although he may not be able to define it, he believes that he knows it when he sees it. However, this is true only to a degree. Excessive quality may not be recognized, especially if there is no means for comparison. Quality to this degree is a waste of money.

After expressing some thoughts on quality at a recent conference, I was asked for comments relative to a furniture manufacturer who was manufacturing honeycombs for a beekeeper for the collection of honey. The manufacturer was making the honeycombs to the same tolerances he used in making furniture and felt that if he could reduce the tolerances, he could increase production and substantially reduce cost. He asked what I thought the effect would be. I suggested that he talk to the bees. If the reduced tolerances did not have any effect on the production of honey, the bees did not care and the extra quality was a waste of money. In effect, I was saying that if the customer did not recognize the higher-quality product, it was unnecessary.

Quality, A Subjective Appraisal

Dr. Edwards Deming is recognized as the father of the Quality Revolution. His statistical quality methods adopted by Japan in the early 1950s have led to its leadership in manufacturing. Dr. Deming said that it is not enough for management to be for quality; they must know what it is and do something about it.[3] However, I have not seen anyplace where he defines quality in a positive, quantitative way.

John Gaspari[4] says that few can give a clear definition of quality other than to say, "I know it when I see it." Other comments regarding quality, based on many articles in periodicals over a period of years, show quality to be defined quite differently by different people, depending on their background. For example, the design engineer may believe quality is meeting the concept objectives; to the test engineer, quality is meeting the design objectives; and on the manufacturing floor, quality usually means producing within the tolerances. As John Gaspari says, most people relate quality to appearance or performance and may frequently measure the quality of the product based on a minor function. For example, a new car may run well and look good but a minor squeak or rattle may destroy the whole concept and cause the customer to feel that the product is poor quality. In each of the examples noted above, quality is related to a perception in the mind of a person. It is a subjective evaluation that may change depending on the role of the observer.

If quality is to be the watchword of the future, we should be able to do a better job of defining it. In fact, we should also recognize that quality is not an issue unto itself. Quality is only one of a number of factors that make for a product perceived in the market place as a good product or one of high quality. Quality is important but only if it makes the product more desirable to the customer.

Value and Performance

A highly desirable product must do what the customer expects it to do. It must also be available when wanted and for an acceptable price. As noted in Chapter 4, this is what we call value, the best combination of performance, availability, and price in the eyes of the customer. Value is often expressed by the symbolism of the three-legged milking stool (Figure 7.1). The three legs, performance, schedule, and cost, must be equal or balanced to assure a flat-surfaced, stable stool that meets the customer's requirement.

In Chapter 5 we pointed out that performance is also called function and that function is the performance the customer expects to get for his money. The value, function, cost relationship is described algebraically in Figure 7.2. In the role of the manufacturer, cost is the sum of material, labor, and variable burden required to produce the product. In the role of the customer, cost becomes price when profit and other elements are added to the cost to get it to the buyer.

Although the formula in Figure 7.2 describes the value relationship, each consumer may have a different value system as described in Chapter 6. Because of this, it is difficult to determine what the customer considers good value. Since this is the case, one approach is to strive for

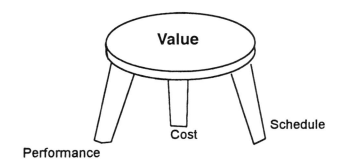

Figure 7.1   The Value Stool

$$Value = \frac{Function}{Cost} \ or \ \frac{Function}{Price}$$

Figure 7.2

best value, the lowest cost to provide the function. If we offer best value or the maximum performance for the lowest price, we would seem to be going in the right direction.

If we go back to the original concept above and analyze the elements of value, performance and price, we find that price is in the mind of the customer. The decision as to whether or not to buy the product depends on his opinion as to whether or not it is a "good deal" or whether he is getting what he wants for a fair price. Whether the price is fair or not is his judgment and may be based on many factors including his ability to pay and his motivation for the product.

I believe the product comes first. The customer is attracted to the product for some reason, usually known only to him. He then makes a decision to buy based on the price; this means determining if he can afford it or if he can get a deal he can afford. For example, if a person wants a new car, he is first attracted by a specific product. He then decides if the price is right. If it is too high, he may go to a second choice to get a better value, the ratio of function to price. If he gets the deal he wants, he then has a good value product. However, if after a brief period, he finds small annoyances such as squeaks and rattles, sticking locks, or loose trim, he may change his mind about a good value because the product is loaded with defects. In other words, it is not performing to expectations. The quality is deemed poor. If we consider these factors and look again at the value equation, it now changes as shown in Figure 7.3.

$$Value = \frac{Function}{Price} = \frac{(Performance + Quality) + Availability}{Price}$$

Figure 7.3

In Figure 7.3 function is defined as performance plus quality when wanted. Performance is the expectation of what the consumer wants the product to do for him. The level of quality reinforces the performance or detracts from it. It must, of course, be available when wanted for a price the consumer is willing to pay or for what he feels is a fair price.

In Figure 7.3 quality is considered to be a component of performance; poor quality detracts from performance, higher quality adds to performance, and in some cases, has been found to increase performance by adding unexpected pleasures not originally considered in the purchase. For this reason, some manufacturers now measure expected benefits or things gone right, (TGR) as well as things gone wrong (TGW).

Higher quality adds to performance and contributes to higher value to the consumer. When quality is highest, performance is maximized; price and availability then become important elements in the equation and introduce just-in-time methods and cost as increasingly important.

Figure 7.4 is an Argus Chart of the value, function (performance), schedule (availability), cost (price) relationship. This plot clearly shows that assuring quality requires the elimination of defects in a product and that to meet customers' expectations it is necessary to ensure quality. Eliminating defects satisfies requirements, and the project will then meet the required expectations for which it was designed. However, it is important to note that quality beyond requirements will not necessarily improve expectations and may throw the balance of function, schedule, and cost out of alignment and not be accepted as a benefit. Thus, it can be an unnecessary expenditure of assets; a waste of money.

Figure 7.5 illustrates this price, performance, availability relationship in terms of value and quality. From this diagram, it can be seen that when cost is added to quality, it becomes value.

Measuring Value

To make specific use of the above information, it is necessary to develop positive measures for each component. In this case, we will be concerned only with value and quality. We believe value is the goal of the consumer and quality is a component of the system. In the chapter on value, we

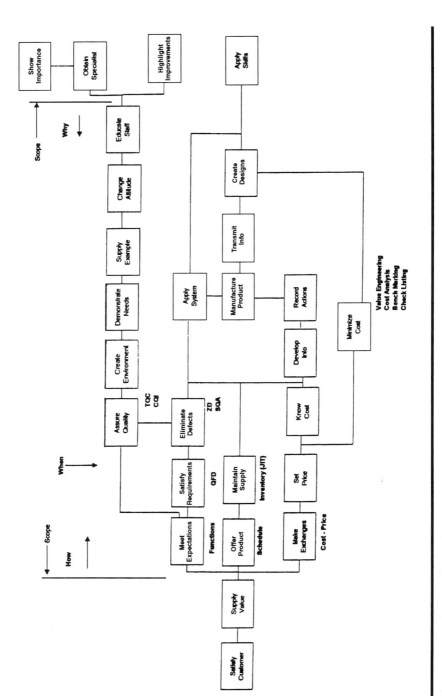

Figure 7.4   Value Quality Relationships — Argus Chart

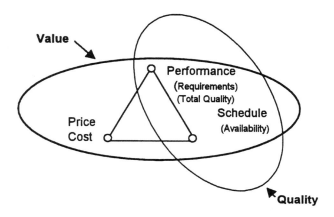

Figure 7.5   Value–Quality Relationship

discussed the factors to be considered in measuring various types of value in both specific and abstract systems. We are now adding a new factor to the system, quality.

Measuring Quality

There are two ways to measure quality. The first and most universal is by the number of defects in a product. The second is the Taguchi Loss Function, discussed in Chapter 2, which is based on the deviation of the product from a target. The loss in Loss Function refers to the loss in profit relative to deviation from baseline performance. This loss can be calculated based on quantitative units or subjective measures set on a prearranged hierarchy. The Taguchi Loss Function is usually applied to manufacturing tolerances; however, there does not appear to be any reason why the same principle cannot be applied to any product.

Count the Defects

We are all familiar with the usual method of determining quality by the number of defects in a product. These defects may be major or minor and the quality level of the product may be determined by the number of each allowed in the product. For example, a product may be deemed acceptable if it has no more than five defects. These may be divided into one major and four minor defects. If it has more than one major defect, it is unacceptable. No major defects indicates good quality. The same would be true for minor defects. In an automobile, a major defect might be stalling, a rough transmission, or high wind noise. Minor defects might be a trunk lid that requires excessive force to close, a loose piece of trim, or a dented wheel cover. The major defects have a direct effect on performance of the vehicle relative to the basic function. The minor defects are annoyances but can be easily remedied.

Unfortunately, in many cases, it is the minor defects that establish the consumer's opinion of the quality level the product. One of the main reasons for this is because of the difficulty in getting them eliminated. Even though the product may have no defects affecting the performance of the product, several minor problems may establish a poor quality level in the opinion of the consumer. At the very least, he has to get them corrected.

These minor defects can also be a major cost problem. At today's rates, a manufacturer cannot tolerate too many defects before his profit disappears. If we consider the cost to repair several defects in a new automobile, we can see how the Loss Function Theory can be applied to any consumer product as a Quality Depreciation Index (QDI).

Quality Depreciation Index (QDI)

Information in a leading consumer magazine indicates that about 3.5 defects per new car purchased by them for test and evaluation appears normal. The defects in the new cars they purchased ranged from one in a $50,000 luxury car to as many as eight in a low-cost, sporty-type vehicle. In general, the number of defects in automobiles has decreased substantially in all brands in the past few years. Although the data show a substantial cost for repairing warranty claims, it is not likely that all defects will disappear for such a complicated product, although it is a goal to strive for.

An analysis of an automobile manufacturer's warranty service reports found that the average time to service a warranty claim was 0.6 hours or 36 minutes. At a rate of $75.00 per hour for dealer cost, this is an average repair cost of $45.00. If only one repair is made on 200,000 cars manufactured, the loss due to warranty repair is $9,000,000 per year. This loss can be charged against quality. If this cost is shown on a simple graph, the cost of deviation from performance can be seen immediately. The graph in Figure 7.6 illustrates the dollar loss resulting from defective products.

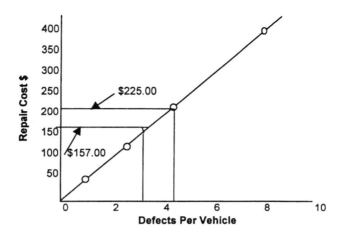

Figure 7.6   Quality Depreciation Index

If we take 3.5 defects as representative and incorporate it into the formula in Figure 7.3, as shown in Figure 7.7, for two $20,000 vehicles, one with 3.5 defects and one with 5 defects, there is a definite decrease in value to the manufacturer in profit and to the consumer in loss of performance and satisfaction. The equation has been multiplied by 1,000 to convert the result to a significant number.

The calculation shows that the vehicle with 3.5 defects has a value measure of −7.2 and the vehicle with 5 defects has a value measure of −11.2, a significant difference.

There is a definite relationship between value and quality. When all the factors are considered, quality is seen as a component of performance as shown in Figure 7.4.

If we consider a quality product as meeting all of the aspects of performance, the consumer receives highest value. If, on the other hand, measurable defects are found in the product, it detracts from value to the manufacturer as well as to the consumer. The manufacturer can measure this loss in value in terms of profit. The consumer, on the other hand, receives a product that does not meet his expectations. If these shortcomings can be corrected promptly and conveniently, the consumer may not regard the loss as a decrease in quality. However, if correction of the shortcomings requires inconvenience, time, and repeated visits for repair he may regard the product as poor quality and never again buy a product made by that manufacturer.

$$Value = \frac{Function}{Price} = \frac{(Performance + Quality) + Availability}{Price}$$

A. Defects 3.5
    Quality Depreciation (QDI) Chart Figure $157.50
    Product price $20,000.00
    Performance   Unit (1)

$$V = \frac{P + (S - 157.50)}{\$20,000} \times 1,000 = -7.8$$

B. Defects 5
    QDI   225
    Product Price $20,000.00
    Performance   Unit (1)

$$V = \frac{P + (S - 225) + A}{\$20,000} \times 1,000 = -11.25$$

Figure 7.7

This analysis is primarily to illustrate the monetary effect of quality on a product. A similar analysis can be made at any stage in the manufacturing of a product. However, retail customers don't usually have the necessary data to make a financial analysis and must rely on methods of comparison based on characteristics they feel are desirable, as illustrated in Chapter 6.

References

1. Significant parts of this chapter are reproduced from, Park, R. J., In value and quality — conflict or complement, *SAVE Proceedings,* Vol. XXVI, Vogl, O. J., Ed., Northbrook, IL, 1991, 95. With permission.
2. From *Business Week,* Quality — how to make it pay, Aug. 8, 1994.
3. Main, J., Under the spell of the quality gurus, *Fortune,* Aug. 18, 30, 1986.
4. Gaspari, J., *I Know It When I See It,* AMACOM American Management Association, New York, NY, 1985, 67.

THE HUMAN ELEMENT

Chapter 8

Communications: A Two-Way Street

Introduction

When Henry II said of Thomas Becket, "Will no one relieve me of this troublesome monk?" he didn't mean to kill him.

A good part of every day is spent conversing with business associates, customers, neighbors, friends, family members, and a wide variety of other people with whom we come in contact. With all of this communicating back and forth you would think we are very good at it. However, we all have examples to prove this assumption wrong. In some cases, misinterpretation of a communication can lead to disaster.

Communication is the art and science of the interchange of ideas and opinions through a system of symbols such as language, signs, and gestures.

As indicated, communication is a two-way street. The purpose of communicating with people is to transmit a message to achieve a mutual understanding between sender and receiver. In a simple discussion regarding the weather, this may be easy. However, in more complex situations, achieving mutual understanding is often difficult. Paradoxically, in instances where it is most important, understanding is most difficult to achieve. If both parties to the conversation or other form of message medium do not understand each other, it is not likely that agreement will result or, in the case of an order, that it will be properly executed.

In any group discussion it is important that to be truly effective each participant should have some understanding of the basic elements of communication. For example, as the discussion progresses, information will be transmitted verbally from one person to another. It is important that each member of the group listens to what is being said. Although each person will hear the same words, each may interpret them differently depending on his or her background. Their general attitude may also be expressed by their body position or facial expressions. Even the tone of voice may indicate enthusiasm or boredom with the subject.

As a group leader trying to create an environment for a successful project result, it is important to recognize these signs and symbols so you may take remedial action to improve group performance if it becomes necessary. It may be necessary to draw some people out to develop better participation or to restrain others who are overpowering the group. If the group has been selected to provide the best knowledge available on the project, it is imperative that as much if not all of the inherent information be available to all participants.

As a group participant it is also your responsibility to participate in the group discussions as freely as possible and to obtain information to allow you to contribute your expertise for a truly effective consensus decision.

The Communications Model

Whether the communication is verbal or written, there must be at least two parties, the sender and the receiver. Most authorities say that to successfully pass a message from the sender to the receiver six steps must take place.

1. Think. The sender has an idea in his mind.
2. Encoding. The sender converts the idea into a communicable form. This may be verbal, written, body or facial signs, art, sounds, or a combination of all.
3. Signals. The transmission of the message.
4. Perceiving. Reception of the message with one or more of the senses.
5. Decoding. The receiver translates the message into understandable form.
6. Understanding. The receiver understands the message as the sender intended.

The communications model in Figure 8.1 is simple. However, if there are any breaks in the chain, the intended message will be altered. The two most likely problem areas are *2. Encoding* and *5. Decoding*. Problems here may be caused by distrust, defensive behavior, misinterpretation, or just plain misunderstanding of the encoded message. For example:

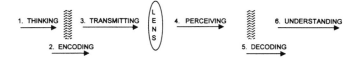

Figure 8.1   Communications Model

Recently a friend, head of a cost department for a major company, was asked to educate a person from a French affiliate in his costing methods. The two worked together for several weeks. Then, by accident, my friend learned that the student did not understand what a die was. A die, in the technical terminology in which it was being used, is a device to aid in forming metal into complex shapes, but in French the word does not have a direct translation. Therefore, the student spent several weeks not having the faintest idea what the teacher was talking about because he was too proud to ask.

Transmitting a message is only half of the system; there must be a receiver, otherwise there would be no reason to send the message. Whomever receives the message interprets it to come to a conclusion as to what the message means. Does it request information, offer assistance, issue a directive, transmit a command, or request some other action? Whatever the message states, how it was transmitted is of the utmost importance.

Was the message transmitted verbally, face to face or by telephone? Was it written, typed, printed, or some other form? Can more be read from how the message was written than from the words? What does the whole system tell you? All of these questions relate to some part of the communication.

Answering these questions involves another aspect of communication, and that is feedback. Not only must there be a receiver for a message there must also be a need to know if the message was understood. In other words the receiver must feed back information to the sender indicating that he understands. In reality the receiver now becomes the sender and reverses the process illustrated in Figure 8.1. This feedback may be verbal or nonverbal. The feedback may indicate that the original receiver understands the message and both sender and receiver are in total agreement. The feedback may also show that the receiver does understand the message, but his nonverbal language indicates that he doesn't like it and is not in agreement. Worst of all is no feedback where the receiver may be like the Frenchman in the previous example. He didn't understand and was too proud to say so.

For these reasons feedback is important. However, in many cases the sender does not receive feedback either because he doesn't look for it or because he doesn't recognize it. As a result, many messages are misunderstood and the desired action never takes place or is not carried out satisfactorily.

Communication involves speaking, writing, listening, reading, verbal acuity, spelling, and several other symbolic characteristics, such as appearance and seating position. Then there is body language, facial expression, and tone of voice as well as inflection.

Nonverbal Communication

Although the participants in the communication effort may not think of them, at least several of the above-mentioned elements are present in every communication. In addition, even though we may not consider many of the more delicate nuances of communication, we are frequently and strongly affected by some of the symbolism mentioned above.

For example, the evening TV news may show the arrest of two men involved in a brawl. Their appearance is unsavory. Their dress is dirty, poorly fitting tee shirts and jeans, unkempt hair, and other slovenly characteristics. However, when they finally get to court their attorney will see to it that they are dressed as professional businessmen to try to create the impression that they are really substantial citizens. Does this work? In general, the answer is yes; we are guided by appearances.

A recent trial involved the testimony of a well-known TV personality. For her court appearance she was dressed in a plain white suit, no jewelry, and a very plain hairstyle. A weekly news magazine commented that she in no way projected the bubbly, effervescent, self-assured person seen on her TV show. The attempt was to create an indication of innocence and of a person who couldn't possibly be involved in such a scheme.

It is a proven fact that appearance creates an immediate impression in people's minds and attorneys try to take advantage of this fact for the benefit of their clients. Should the person be dressed up or dressed down? It depends upon the image you are trying to present.

This is one form of a subtle powerful symbolic communication. However, there are others. Body language is one that frequently is not so subtle. If your audience is yawning and finding it difficult to stay awake, your presentation is in trouble. However, it may also be in trouble if a person crosses his arms over his chest or his hands over his mouth, basically he is signaling that he has heard enough.

Then there is the impression that may be created in the room layout. Who is sitting at the head of the table or who has the largest chair? Even these small acts may be signaling who the power is in the room and the one you have to convince of the benefits of your proposal.

So what does this mean? It means that speaking and writing are only two ways to transmit information. Although they may be very important, they may be strongly affected by many more subtle forms that transcend your message. For example, certain signals sent to the audience by one in the power position may cause other members to withhold comment or even to vote in a certain way toward the proposal.

We should be aware that these forms of communication can be very powerful and should not be overlooked when making a presentation. Although we may not be able to control them, we certainly should be aware of the signs so that we can avoid any unwanted conflicts.

A major advertising agency considers these nonverbal communication symbols to be of the utmost importance in creating TV commercials and cites the Mehrabian Study as an indication of their importance. The study claimed that of all means of communication facial expression accounted for 55% of the overall effect, tone of voice 33% and words only 7%. This clearly shows the effect of symbolism in communication.

The fact that vocal communication appears to have only a part of the effect on a communication does not mean it should be neglected. We should make every effort to build a suitable vocabulary for all occasions and learn to articulate clearly, eliminate as many regional speech patterns as possible, and learn to speak with adequate volume in order to be heard.

Listening

It is true, of course, that a major part of our communication is oral. We spend a large percentage of our time speaking with other people either to transmit information of some kind or to request information for some purpose. However, we give very little thought to the fact that an equal amount of time is spent listening to an oral transmission from another person. There really would be no reason to speak if there was no one to receive our message or request.

There are several traditionally accepted steps in the learning process as follows:

- Recognition of knowledge
- Assimilation of knowledge
- Application of knowledge
- Generation of new ideas
  Combining ideas
  Finding new ideas

	Listening	*Speaking*	*Reading*	*Writing*
Learned	1st	2nd	3rd	4th
Used	(45%)	(30%)	(16%)	(9%)
Taught	Least	Next least	Next most	Most

Figure 8.2   Communication Skills — Learned — Used — Taught

We cannot understand, assimilate, or use knowledge we have not seen or heard. So poor or ineffective listening may very well put an end to the learning process before it actually begins.

Practical research over the past 30 years has confirmed that approximately 70% of our conscious waking day is spent in communication. That amounts to 7 out of 10 hours of our waking time. Numerous studies have confirmed that out of these 10 hours, approximately 1 hour is spent in writing, 2 hours in reading, and 3 hours in talking. However, 45%, or over 4 hours, of the time is spent in listening.[1]

Interestingly, the educational system spends a considerable amount of money teaching children how to read and write and sometimes how to speak, but there doesn't seem to be any attempt to teach a person how to listen effectively. Charles Schulz, the creator of "Peanuts," often uses communication as the subject for his comic strip and expresses the general attitude toward the subject. In one case, he has his characters discussing listening, the fourth step in the communication model shown in Figure 8.1. In the discussion Charlie Brown says, "Listening is not a bystander sport; you have to try to be a better listener."

Dr. Ralph G. Nichols[2] is widely considered to be the pioneer in scientific research on listening, and he states that if we could subject ourselves and our children to even 10 minutes of uncomfortably tough listening once a week for 12 years, we would not have the present great washout at the university freshman level.

Listening is important, as shown in Figure 8.2. Although a major part of our time is spent communicating, little or no time is spent on teaching people how to listen.

Dr. Nichols has identified what he considers to be the 10 worst listening habits of the American people. I have listed them here. In Figure 8.3 I have combined Dr. Nichols' list with that of Dr. Lymen K. Steil. Dr. Steil is Dr. Nichol's successor at the University of Minnesota.

1. *Calling the subject uninteresting.* Tune the speaker out simply based on the subject title, for example, communication. Rather than turning the speaker off, you might listen to see if the speaker

Bad Listener	Improve	Good Listener
1. Calling the subject uninteresting. Tunes out dry subject	Find areas of interest.	Ask, "What's in it for me?"
2. Criticizing speaker's delivery. Tunes out if poor delivery	Judge content, not delivery.	Judges content, ignores delivery.
3. Getting overstimulated. Tends to enter into argument	Hold your fire.	Don't judge until comprehension is complete.
4. Listen only for facts. Listens for facts	Listen for ideas.	Listen for central themes.
5. Trying to outline presentation. Takes extensive notes	Be flexible.	Take fewer notes. Use different systems depending on speaker.
6. Fakes attention. Shows no energy output	Work at listening.	Works hard. Is physically active.
7. Tolerating or creating distractions. Distracts easily	Resist distractions.	Fights or avoids distractions. Concentrates.
8. Evades difficult material. Seeks only light recreational material	Exercise your mind.	Uses heavier material for mind exercise.
9. Emotion laden words distract listener. Reacts to emotional words	Keep your mind open.	Interprets color words.
10. Wasting the differential time between speech and thought. Tends to daydream with slow speaker	Capitalize on fact that thought is faster than speech.	Challenge, anticipate, summarize, weigh the evidence.

Figure 8.3   Ten Listening Habits and How to Improve Your Listening Ability. (Adapted from Nichols, R. G., *Listening Is Good Business*, Industrial College of the Armed Forces, Washington, D.C., Pub. No. L 65-13, 1964. With permission.)

has anything to say that you might not have heard before, something that you can use for your benefit.

2. *Criticizing the speaker's delivery.* We may begin to criticize the speaker based on the tone of his or her voice or accent — if not voice, then appearance, style, or perhaps a disturbing trait. However, the message is the important element. If the speaker has something important to say, you may become so interested that you will no longer notice his or her appearance.

3. *Getting over-stimulated.* This is a matter of disagreeing with the speaker and spending your time developing a rebuttal. A good practice is to hear him or her out before you pass judgment.

4. *Listen only for facts.* Try to get the main idea out of the discussion, then the facts can be tied into the principles.

5. *Trying to make an outline out of the message.* Only a few speakers follow an outline. In addition, a carefully inscribed outline is usually of little use several months later. The most effective way to take notes seems to be the facts versus principles system. Listen for the principles and then jot several facts to support it. A quick review of the principles will be supported by the facts.

6. *Faking attention to the speaker.* Good listening is not a relaxed and passive situation. It is hard work, and the audience should look and act alert. They should be trying to learn something.

7. *Tolerating or creating distractions.* Speak up if you can't hear the speaker and ask him or her to talk louder; it is the speaker's obligation to speak to be heard. If others are creating a distraction, advise them of the problem.

8. *Avoiding difficult material.* Make a practice of listening to a lecture on scientific material or listen to TV programs such as "Meet the Press," rather than simple comedy or detective programs.

9. *Letting emotion-laden words get between you and the speaker.* It is a fact that a single word may have such an emotional effect that it will cause some listener to tune the speaker out. Even in this day, one of these words is "evolution." There are many others that vary according to time and place but are sure to cause certain members of the audience to tune out or become very belligerent.

10. *Wasting the differential between speech and thought speed.* Speech speed runs about 100 to 125 words per minute, but it is very easy for a person to absorb at least 450 words per minute. If a speaker is cruising too slowly, the listener will drift off into his own thoughts and forget about listening.

How to Improve Your Listening

Dr. Nichols recommends three things that you can do to improve your listening ability and eliminate or reduce your bad listening habits.

1. *Anticipate the speaker's next point.* Try to guess the point the speaker is trying to make.
2. *Identify what the speaker has for evidence.* Does he or she have the support for the points being made?
3. *Concentrate.* The differential between speech and thinking speed allows you to concentrate, then analyze the result before concentrating on the speaker again.

Although people with hearing difficulties have trouble listening, they are often good listeners because they try harder. If your listening can be improved by improving your hearing, do something about it. Move closer to speakers, ask them to speak louder. Read their lips etc.

Meetings

No discussion of communication would be complete without some comments on meetings. David Hon,[3] National Training Manager for the American Heart Association, says that there are four reasons for meetings and identifies the types as follows: Muster, Medium, Assembly, and Task-Oriented.

The Muster

It is necessary to keep the channels of communication open in any organization, and one way to do that is to call Muster-type meetings. The Muster is simply a get-together where people can meet each other, renew acquaintance, and make contacts for future dealings. Examples of Musters are some weekly departmental meetings, monthly club meetings, and cocktail parties. They are informal in structure, and there is little responsibility on the part of the attendee other than to be there. In addition, it is only necessary for the person who called the meeting to see that it is properly organized to suit the occasion.

The Medium

This type of meeting is designed to transfer information and usually involves speeches and/or briefings. Examples of the Medium-type meetings would be program reviews, sales training seminars, monthly technical society meetings, and on a larger scale, symposiums and national conventions. Very frequently the Medium can be a combination of the Muster and the Medium. Many attend to learn; others come to renew acquaintance and to network with others in the hope of obtaining new business. The Medium frequently requires substantial effort to organize a program and supporting activities, accommodations, etc.

The Assembly

The Assembly can also be a combination of the Muster and the Medium; however, some result is usually expected. An Assembly is usually called to approve or reject proposals made by other groups. Some examples of Assemblies are Board of Directors meetings, general memberships in organizations, stockholders' meetings, and periodic meetings to check and support progress on project developments. A good example of an assembly would be the Value Engineering workshop meeting to present ideas and recommendations to management.

The assembly is a preplanned program with an agenda identifying subjects to be brought before the group and the result expected. The outcome of these meetings is usually to begin or end the work of other groups such as task-oriented groups.

Task-oriented Meetings

The Task-oriented meeting differs from the others in that it is goal oriented and some result is expected from the session. A successful Task-oriented meeting must create something, and it accomplishes this feat by making use of the group mind. The Task-oriented meeting sets goals, analyzes all available information, identifies and solves problems, makes decisions, and plans future actions to accomplish objectives. This is a crucial type of meeting, totally result oriented. Usually these meetings are made up of a small group of individuals with diversified backgrounds to provide a broad understanding of all facets of the project. Task-oriented meetings, how to organize them and how to guide them, will be discussed under Teams and Teamwork, Chapter 10.

Discussion

A situation that may arise in any meeting is the discussion. Discussion is essential to the Task-oriented meeting; however, discussion should be kept to a minimum in any meeting. This is not to say that discussions should be discouraged but rather that they be kept under control. Frequently, discussions will absorb the meeting, and although everyone will agree that there was a good discussion of the subject, nothing may have been accomplished. *Robert's Rules of Order*[4] was devised to aid in controlling the course and flow of meetings and offer the best opportunity for accomplishment.

It is often very difficult to control discussion in a meeting of volunteers. The volunteer feels that he or she is contributing time for the cause but wants everyone to know how much work he or she is doing, and this leads to discussion. In most cases, it is necessary to tolerate some discussion to show that the person is appreciated, but unlimited discussion should be avoided at all cost.

Note Taking

In any meeting there may be a need to take notes for future reference. This need may arise more frequently in the Assembly and Task-oriented meetings. It is a fact that you can't be listening very effectively if you are busy taking notes, so over the years I have found a simple technique described by Dr. Nichols as very useful. However, I have modified the technique somewhat for clarity and efficiency.

The method is simple. Divide a sheet of paper or a page in your notebook into two columns. Head the left column *Principles* and the right column *Facts*. Now sit back and listen for the *Principles* and when necessary add a few supporting *Facts*. It is not necessary to add too many facts, because the principles are the key. Reviewing them will call the facts to mind and may even cause you to think of a few that hadn't been mentioned. If you have become skilled at function definition discussed in Chapter 5, you can use the two-word function definitions to define both the principles and facts and find you have clarified the whole process of note taking. The only problem is that it is not easy to identify principles. Part of the problem is that speakers often digress widely from the subject and tend to obscure the principles with facts and other useless information. However, practice makes perfect and practice in identifying the principles will improve your listening habits and knowledge.

Meeting Productivity

A noted humorist says that there are really two kinds of work in most modern companies

1. Taking phone messages for people who are in meetings.
2. Going to meetings.

He says going to meetings is better because that's where the prestige is.

Although this may make you smile, it is basically true. Going to meetings provides visibility and indicates some level of importance by the fact that someone felt that you could make some type of contribution to the effort involved. At one time I was told by a person who seemed to spend his entire life in meetings that he went to all meetings whether or not he had any interest or apparent reason because he didn't want anyone to forget to invite him to a meeting that he did want to attend. This seems like a poor reason to attend a meeting and makes one wonder if he really did any productive work. It also makes me wonder how many others like him make a life out of attending meetings and what can be done about it. To eliminate this type of problem, criteria should be established for attendance at meetings and attendees chosen based on their contribution toward a productive group.

Taking phone messages for people who are attending meetings has now been replaced in most companies by some form of voice mail or answering machines. In the past when someone answered the phone, it did disrupt the answerer's routine, but it usually identified when your party was expected to return. However, the automatic machine response often has a serious drawback. Since it is not possible to ask the machine a question, you are frequently left in a quandary because there is no way to tell if the person is away for an hour, a day, for some extended period of time, perhaps even on vacation. In studies I have conducted for several companies I have found that this can cause serious productivity problems. In one case, 25% of work time was lost in a specific laboratory because of this condition. If electronic message centers are to be used, I would recommend that people be taught how to use them and clearly cite the consequences for not using them properly.

In any meeting listening is important. However, in the task-oriented meeting group, listening is of utmost importance. In the first place information must be collected for meeting deliberations. In many task-oriented meetings the required information is collected before the meeting, but it must be reviewed and analyzed to assure all participants understand the information that is available or to point out shortcomings so that it can be obtained. In this stage at a multidiscipline meeting there is not only an opportunity to learn about aspects of the job outside the specific area of your expertise but a requirement to do so.

Each person in the group will be from a different area of the operation and will have different goals and objectives. Each person's values will be different from another's. It is important to hear other views so that the best decision can be made for the best final decision and the preparation of recommendations.

Thus good listening habits will not only have a more productive effect on the meeting, they will also assure the best recommendation after considering all of the factors involved.

The productivity of meetings was clearly pointed out following a discussion with a colleague regarding note taking. I suggested the facts vs. principles method. Some time later I was in my office and a friend came into my office and simply threw a pad on my desk and said, "It really works!" I was stunned for a few seconds and then I noticed that the page had been divided into two sections labeled *Principles* and *Facts*; otherwise the page was blank. He had just attended a two-hour meeting where absolutely nothing was accomplished. Don't waste your valuable time this way. Organize your meetings properly to increase interest and accomplishment.

References

1. Kaminski, S. H., Personal communication, 1998.
2. Nichols, R. G., Listening is good business, *Industrial College of the Armed Forces Seminar Publication L65-13*, 1965. With permission.
3. Hon, D., *Meetings That Matter*, John Wiley & Sons, Inc., New York, NY, 1980.
4. Robert, H. M., *Robert's Rules Of Order*, 3rd ed., Pyramid Books, New York, NY, 1968.

Chapter 9

Motivation: Different Things Move Different People

The Life Cycle

Change

People do things for their own reasons. This is basically to obtain something they want or to avoid something they don't want. There are more complex variations on this theme, but if we understand these basic reasons we can make a marked improvement in ourselves and in our dealing with other people.

Life is a continually changing condition. The seasons change. In the spring we look forward to the warm days of summer. The flowers spring to life and warm weather draws us outdoors to enjoy the weather and the freedom it gives. Then it begins to change as the flowers wither, the leaves turn color and fall, to set an entirely new pattern of life. We move indoors to a great extent and prepare for winter, but as winter approaches, we look forward to spring and summer again. It is a constantly changing cycle that we understand, look forward to, and are prepared to experience.

Since we have adapted to these changes, why is it so difficult to accept new ideas, methods, or procedures? Probably because they lead us into the unknown, and we do not know what will happen. John Dewey said, "All thinking involves risk. Certainty cannot be guaranteed in advance.

The invasion of the unknown is of the nature of an adventure." We cannot be sure of the outcome. We cannot be sure whether the result will be a success or failure. Although it has been said that failure tests our resolve, it motivates us to learn more and forces us to recognize our weaknesses. Fear of failure has been identified as one of the greatest reasons people resist change.

However, although we accept change in the seasons as inevitable, we don't seem to realize that all change is inevitable; it comes on whether we want to accept it or not. We can either react to change or learn how to make it work for our benefit.

Will the change cause failure or success? Unfortunately, we generally worry about the failure rather than the success. This signals danger, which results in anxiety and fear, which causes stress. Over time we have come to recognize that change has a distinct effect on our well-being. So we may resist any action with the hope that the change will go away. Unfortunately, it rarely does. A cardinal rule of life which many of us have yet to learn is that it is not the events or changes but how we view them that is the problem.

Stress

Why do we resist change so energetically? Although change can have a beneficial effect on our lives, we usually see change as the source of problems. It forces us to react to the effect of the change, and this brings on stress. Stress is a perception of threat or discomfort that activates the organism and creates an uncomfortable and dangerous physical condition. Interestingly, it is not the change or the result of the change that affects us but the perception of the effect the change will have on us.

If you are a plant manager and finally have the plant running like clockwork and you have just learned that a new product will now be added to the system, everything will have to be changed, and this means problems. If a new manager has just taken over your department, this means change in the way you have been doing things and again, problems. If you have just been promoted and transferred to a new location, along with the benefits will come problems, not only for you but for your family. If a product in the design–development cycle is over the cost target something must be done; however, taking the time to make changes will certainly cause lost time, require manpower, and cause a host of other problems. Very little thought is given to the possibility that these problems may not come to pass and the product may be changed for the better. The whole series of problems is due to our perception of the task and causes us to resist the new conditions as vigorously as possible. This negative approach to change is all the result of attitude.

So change is perceived to create problems, but consider the opportunities. The new product may make it possible to increase the productivity of your operation by making better use of some of your new machines. The new department manager may need help in understanding the operation and give you a chance to incorporate ideas you have been trying to make for some time. The new job may take you to a new location offering advantages not available at your present location.

It is certainly true we must constantly adapt to our surroundings and general conditions of life. However, even though we must constantly make changes to our lifestyle, these changes do not always create problems; they may frequently open the way to substantial benefits. Do the benefits outweigh the problems or are we unwilling to consider the entire situation? The more negative the perception of the change the greater the stress will be, and this may create a serious health problem.

Change and Stress

Shortly after World War II, Harold G. Wolff[1] at the Cornell Medical Center in New York conducted a series of experiments related to the effect of a constant series of changes on the individual. As a result of these experiments, he emphasized that the health of a person was intimately related to the demands made on him by the environment. From this theory, one of Dr. Wolff's colleagues at Cornell, Dr. Thomas H. Holmes,[2] developed the idea that it was not the individual change but the rate of change in a person's life that could be the greatest environmental factor.

Shortly thereafter and now at the University of Washington School of Medicine, Dr. Holmes teamed up with a psychiatrist, Dr. Richard Rahe, and after extensive research with over 5,000 people in all walks of life in the United States and Japan, developed a Life-Change Units Scale and questionnaire, shown in Figure 9.1. Subsequently, this scale has been verified with the history of thousands of people and clearly indicates that the greater your score on the scale the more likely you are to develop a serious illness within the next year. So it is a proven fact that change can make you ill.

The latest indications are that if you score 300 or more on the scale you have a 90% chance of developing a serious illness within the next year. If you score between 150 and 299, your chances drop to 50% and if you score 180 or less, your chances of developing a serious illness drop to 30%. So we know that change affects health even though the changes may ultimately have a beneficial effect on our lives.

Life- Change Units Scale

ITEM	VALUE	EVENT
1	100	Death of a spouse
2	73	Divorce
3	65	Marital separation
4	63	Jail term
5	63	Death of close family member
6	53	Personal injury or loss
7	50	Marriage
8	47	Fired at work
9	45	Marital reconciliation
10	45	Retirement
11	44	Change in health of family member
12	40	Pregnancy
13	39	Sex difficulties
14	39	Gain of new family member
15	39	Business readjustment
16	38	Change in financial state
17	37	Death of close friend
18	36	Change to different line of work
19	35	Change in number of arguments with spouse
20	31	Mortgage over $10,000
21	30	Foreclosure of mortgage or loan
22	29	Change in responsibilities at work
23	29	Son or daughter leaving home
24	29	Trouble with in-laws
25	28	Outstanding personal achievement
26	26	Wife begins or stops work
27	26	Begin or end school
28	25	Change in living conditions
29	24	Revision in personal habits
30	23	Trouble with boss
31	20	Change in work hours or conditions
32	20	Change in residence
33	20	Change in schools
34	19	Change in recreation
35	19	Change in church activities
36	18	Change in social activities
37	17	Mortgage or loan less than $10,000
38	16	Change in sleeping habits
39	15	Change in number of family get-togethers
40	15	Change in eating habits
41	13	Vacation
42	12	Christmas
43	11	Minor violations of the law
		Total score for 12 month

Figure 9.1   Life Change Units Scale. (Reprinted from Holmes, T. H., The social readjustment scale, *Journal of Psychosomatic Research II*, 213–218, Elsevier Science, Oxford, England, 1967. With permission.)

We know that change and stress can create genuine physical problems. However, there are ways to deal with the problem so that the effect will be less and thereby open the door to acceptance of change and the opportunities it frequently offers.

Dealing with Stress and Change

Medical research has defined three levels or components of the stress reaction:

1. The environment
2. The appraisal and evaluation of the environment
3. The reaction of emotional and physiological arousal

Belief that negative consequences will follow from events causes stress. Once this negative appraisal is made, stress follows automatically. Reducing stress means the elimination of the changes that occur during the last phase of the stress reaction. This can be done to some degree at any of three levels.

We could alter the environment by changing jobs, living conditions, or lifestyle. In most cases, these may be very difficult to do. However, it is usually possible to make some changes that may prove beneficial.

A second step could be the emotional and psychological response to stress. The evaluation of the changes can be weakened by the use of tranquilizers or other drugs. However, the drugs may bring more problems than they eliminate. It is also possible to use mediation and self-hypnosis methods.

The third method to prevent stress is to alter beliefs, assumptions, and ineffective ways of thinking that make us vulnerable to stress. Changing our outlook on some part of our life affects attitudes and can be difficult; however, making the required adjustments is entirely within our control and depends on no one but ourselves. We are the only ones who can do this.

In simplified form, Dr. Robert Elliot, a cardiologist at the University of Nebraska, says there are two rules for dealing with stress. "Rule No 1 is, don't sweat the small stuff. Rule No 2 is, it's all small stuff and if you can't fight and you can't flee, flow."

Change of Attitude — It has been recognized that changing the way a successful person attacks a problem can be stressful. So we looked at the potential effect that changing a person's attitude has on them during the course of a workshop. Since we deal with engineers extensively, we

"OUR STAFF INCENTIVE FOR CONTINUED
ENGINEERING EXCELLENCE IS SIMPLE.....
WHEN A BUMPER PROPOSAL FAILS THE CRASH
TEST, WE BARRIER IMPACT THE DESIGNER !!!"

HERLITZ '72

Figure 9.2   Engineer's Priorities

looked at the effect of change on an engineer and found that the engineer sets his priorities as follows:

1.   If the product fails to work, he's a dead duck.
2.   If the product is late, management will provide all the necessary support to expedite it.
3.   If the product is over the cost target, management will set up a cost reduction committee to help reduce the cost, and the engineer who put all the cost into the design will be appointed chairman.

What would you do? The cartoon in Figure 9.2 clearly illustrates the action.

Why We Act — The basic motivational pattern illustrated in Figure 9.3 shows that motivation comes from within the individual. People do things for their own reasons. If they see a change as offering a reward, they may take action to achieve the benefit. However, if they perceive the project as producing a product change that has a potential for failure, they will do everything in their power to forestall the need for any action on their part.

Fear of failure is recognized as one of the greatest barriers to change and hence progress. I have seen projects fail that offered the benefit of substantial cost and weight saving, in addition to improving the product. The recommendations were accepted and many thousands of dollars were spent on development models for test and evaluation. The models had been received and tests begun when a high-level management authority

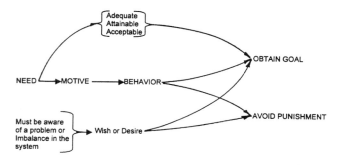

Figure 9.3   Basic Motivation Pattern

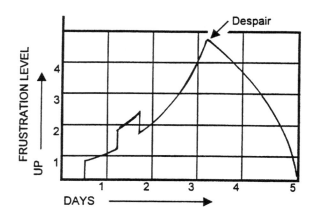

Figure 9.4   Frustration Index

decided that the changes might not work and suggested taking the best parts of the recommendation and modifying the existing product. As a result, nothing was accomplished.

Frustration Index — As a result of our analysis of the overall picture produced by change and stress, we developed a Frustration Index, shown in Figure 9.4. This chart illustrates the stress pattern generally experienced by people as they go through a new method of problem solving. It is based on a five-day Value Engineering workshop program that has been proven successful in thousands of projects.

The chart indicates that in most cases, half of the time spent on a project is spent in precisely defining the problem. This is a particularly stressful period for most people because they are anxious to solve the problem even though experience has shown that they really don't know what the problem is. It is additionally stressful because in this initial period the project with which they are thoroughly familiar is being dissected into a chaotic mess. However, if the process is followed, success will be the result.

Our intention is not to eliminate the stress but to illustrate that it is a normal pattern and to change the perception for a more effective and relaxed approach.

Habits and Attitudes

A habit is an acquired or developed mode of behavior that has become nearly or completely voluntary, a custom, practice, or way. An attitude is a behavior representative of feeling or conviction, A persistent disposition to act either positively or negatively toward a person, group, object, situation, or value.

It is very difficult to break or change our habits because our attitude reinforces our behavior. Although habits may force us to do the same old thing the same old way, they also have their good points. If we had to think about how to tie our shoes every morning, we would probably be late for work on most days. Habits do not have to be simple acts. Athletes frequently program themselves to perform certain complex programs that they perform over and over. So habits can save us time and mental effort. On the other hand, our habits frequently make it very difficult for us to be creative and to think up new ways to do things. Unfortunately, the effort required to break down our habits and attitudes can create stress.

During the course of working with many people in many different industries over a long period of time I have come to believe that it is important to be careful of the recommendations of "experts." They develop habitual ways of thinking, and their attitude helps to keep them doing the same thing the same way. Breaking down this stereotyped thinking can be difficult but important. Sales people say that doing the same thing over and over and expecting different results is insanity. Frank Lloyd Wright has been quoted as saying, "An expert is a man who has stopped thinking."

It has been my experience that the typical reaction when showing a new idea to an expert is as follows: "That sounds like a great idea. Let me tell you what's wrong with it." So you listen and make notes and when the expert has listed 10 or 12 problems "off the top of his head" pick one and say, "This is great. Now that we know the problems, how can we solve this one?" Before the discussion is over you may have him or her working with you to make your idea work.

"Heavier than air flying machines are impossible."
Lord Kelvin, President, Royal Society, 1895

"Everything that can be invented has been invented."
Charles H. Duel, Director U.S. Patent Office, 1899

"Sensible and responsible women do not want to vote."
Grover Cleveland, President, United States, 1905

"There is no likelihood that men can ever tap the power of the atom."
Robert Millikan, Novel Prize in Physics, 1923

"There is not the slightest indication that nuclear energy will ever by obtainable. It would mean that the atom would have to be shattered at will."
Albert Einstein, 1932

"That is the biggest fool thing we have ever done.... The bomb will never go off, and I speak as an expert in explosives."
Adm. William Leahy, Advisor to President Harry Truman, 1945

Figure 9.5   Advice from the Experts

Some comments over a number of years and in different generations on several issues shows that although times have changed the human reaction hasn't. Now that we have had a chance to look back on some of the responses in Figure 9.5, they seem ridiculous. However, we must remember that the responses are from those who knew what they were talking about at the time. They were the experts; who could question them?

We must learn to reply to the new idea by saying, "That sounds like a good idea. I see some problems, but let us see if we can make it work."

Nature of Man — The chart in Figure 9.6 illustrates in simple diagrammatic form the nature of man as seen by psychologists. Basically, the human being is made up of a body and mind. The body receives input into the mind through the five senses, sight, sound, touch, smell, and taste, and translates this information in the mind based on an evaluation analyzed by the emotions and intellect. The resultant decision is transmitted to the will, which takes the action agreed upon. In many cases, this action may be no action at all. The memory is part of the intellect and includes all of a person's past history and experience, including habits and attitudes. The emotions are based largely on imagination.

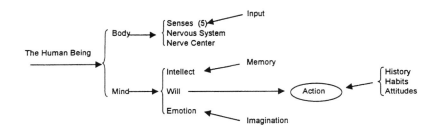

Figure 9.6   The Nature of Man

From this chart, it is easy to see how our habits and attitudes affect our actions. No matter what a person is told, he will process the information in accordance with his experiences and emotions. If we refer to the chart in Figure 8.1 in Chapter 8, Communication, we will see that after encoding, the information is transmitted through a "lens." How it is perceived on the receiving side will determine how it is decoded. In other words, we see things depending on how our lens has been ground, and that is the result of our history and experiences. The result is that we cannot make a person do anything that doesn't fit in with his or her psyche. It is extremely important to realize this fact when working with a team of people of diversified backgrounds. You may need their information but until they see the project as beneficial in terms of their experiences, there may be strong disagreement among the group, and it will be necessary to break down these mental constraints to achieve successful cooperation.

JOHARI Window[3] — In any group made up of people of different backgrounds and experiences there will be a difference of opinion and knowledge. In the initial stages of team development these differences will become readily apparent. There will be things about the project that one person will know and another person may never have heard about. There may also be people who know the same thing but hold widely different opinions about it. In addition, you may be transmitting information of which you are not aware by your attitude, actions, tone of voice, or other nonverbal communication. These differences occur in every group composed of people with widely diversified backgrounds. However, in order to successfully solve a problem or accomplish a task, it is necessary for all participants to understand all sides of the project so that a consensus can be developed for a recommendation.

The Johari Window (Figure 9.7) can be used for self assessment or for dealing with the dynamics of a group. It is a way to describe personality differences in diagrammatic form, and an understanding of the dynamics

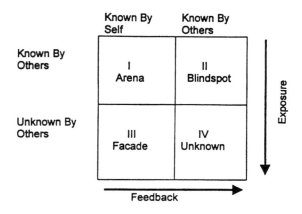

Figure 9.7 Johari Window

of the window can be useful in organizing group operations as well as in understanding how an individual can react in a group environment.

The objective of any team operation is to increase the knowledge of everyone in the group relative to the task so that they can make effective contributions. In the early stages of group development there is always one person who seems to know all about the project. He or she is very knowledgeable, and others appear to know very little. Relative to the Johari Window this means at least one person has a large Façade. This tends to make others restrained because they do not want to appear ignorant. However, as time progresses it becomes apparent that other people know things about the project that he or she does not; as a result their Façade begins to grow. However, by adding their knowledge to the pool the group begins to increase the Arena, the area where everyone begins to contribute in accordance with their capability. As the Arena increases, the window begins to change shape, decreasing the size of the Unknown, or things that no one knows.

The Johari Window also shows how the process of integrating a group of people with different backgrounds can cause a stressful situation. This can be especially true in area II, Blindspot. Here, one person may have little or no knowledge with regard to the project, but the other members of the group may be thoroughly familiar with it. The reverse may be true in area I, Arena. One person may be the knowledgeable person in a specific area of the project, and the group must rely upon him or her for any information in that phase. The result can be a stressful situation because no one wants to appear ignorant even though their knowledge may be limited to only a small degree.

The Johari Window concept can be an aid in organizing a team or in dealing with conflicts within a group to relieve tensions and to increase participation from the more reluctant members of the group.

Summary

When working to solve a complex problem by using a new and different method, we are asking you to change the way you have always looked at a problem. This means you must struggle to break down the constraints to visualization that you have developed over a lifetime. This puts you under stress. In the case of a single workshop, it will be only for a few days, then you can go back to the old way of doing things. We hope that will not be the case for all of you. We think the breaking down of mental constraints can have a beneficial effect on your professional performance if you will try to apply them.

There usually has not been great effort put into teaching engineers and other technical people nontechnical subjects. However, after a while many begin to realize that things do not always work as smoothly as we would like. We see that everyone is not moved by the same rewards or benefits. Understanding some of the factors that cause negative reactions to new ideas and why they are a part of human nature may make it possible for you to present your case in a way that will be accepted rather than resisted and bring benefit to all involved in the transaction.

Why does changing the way you do things create stress? You might fail! On the other hand, you might develop a new method or theory. So we have a dilemma. Success or failure?

In a recent project, after careful evaluation of a product, the team decided to recommend the elimination of two parts that were designed for a specific purpose but were not being used as proposed by either repair people or customers. They decided to modify an existing component to satisfy the basic requirement. The potential piece cost benefit was substantial, and over a three-year period, the saving would amount to over 7 million dollars.

The recommendation was received quite favorably by management. However, the engineer responsible for the product resisted the recommendation vigorously, even though the parts were not being used as intended and the suggested modifications overcame a potentially damaging condition. Why the resistance?

As long as the parts were on the product, there was the appearance that the company had considered the possibility of damage or injury. Once the parts were removed this visible indication was eliminated even though a better arrangement was made by the recommendation. This put the engineer in danger of criticism, and he was protecting himself from something that not only may never happen but could be made better for less cost by a more robust design of an existing component.

This is a normal reaction to a change. We must begin to ask ourselves how we can do something better. Rather than saying we can't do that, we must begin to say, "Let us think of at least one good reason why we should make this change."

References

1. Toffler, A., *Future Shock*, Bantam Books, New York, NY, 1971, 327.
2. Holmes T. H., The social readjustment scale, *Journal of Psychosomatic Research*, Vol. 11, 213, 1967.
3. Picraux, F. J., Chrysler Corporation Value Engineering Workshops, Personal communication, 1972 to 1975.

Chapter 10

Teams and Teamwork: A Synthesized Knowledge Group

Teams for Overall Understanding

It may still be possible for one person to know all that is necessary to conceive, design, and manufacture a product for today's marketplace. However, it will probably be a simple product because our complex society and technology have tended to make us specialists in a limited area of activity. This specialization has caused us to compartmentalize our operations and to a large degree our thinking. The more complex the organization the more it has been fragmented into autonomous units that deal in only a small part of the overall process.

In 1927 Charles A. Lindbergh[1] and Donald Hall, Chief Engineer of Ryan Airlines, sat on Coronado Beach in San Diego and established the basic criteria for the Spirit of St. Louis. Two people knew all that was needed to develop an advanced product that even today clearly shows their creative thinking. Lindbergh established the requirements, Hall provided the technical knowledge, and the thirteen Ryan supervisors and employees provided the understanding, know-how, and enthusiasm to develop a product that was designed, tested, and won everlasting glory for Mr. Lindbergh, all within 13 weeks. The product was designed to perform a specific function for a specific cost target. There was no communication problem. There was no cost problem since $15,000 was all they had for the entire project, and there was no timing problem. Any delay was unthinkable.

Consider the design of an advanced aircraft today. The cost in men and materials is almost beyond comprehension. Hundreds of thousands of people in dozens of industries in several states and a number of foreign countries work in vast industrial complexes for years before the product takes to the air. A recent technical magazine article pointed out that 4,000 subcontractors in 42 states and several foreign countries are required to produce a present-day fighter plane. Every day we see articles in business and technical journals and other publications indicating that developing the next generation of large airliners will cost over several billion dollars and will take about 5 years.

The largest manufacturing activity in the world is the automobile industry. Approximately 50 million new vehicles are produced each year, and the industry affects people in every corner of the U.S. and in many countries worldwide and is still growing. In addition, the industry is working hard to speed up the new product development cycle.

Is there any wonder that we hear the word teamwork more and more in industry? It is the only way to organize the necessary knowledge to simultaneously consider all of the factors involved in designing and developing a new product. However, it must be clearly recognized that just any group of people is not a team.

Population and Ability

A statistical finding applied to bureaucratic systems such as governments and large companies shows that work capacity of a population sample will follow a typical distribution curve similar to Figure 10.1. The curve shows an expected distribution of athletic ability from low to Olympic standard which can be readily measured.[2]

The curve shows two countries, Country A with a population of 250 million people and Country B with a population of 25 million people. Country A can be expected to produce a large number of Olympic-standard athletes in many fields, and to develop successful leaders. Country B, on the other hand, will only be able to produce an occasional Olympic-standard athlete, and it will never be able to predict when they will arise or in what field they may be.

This hypothesis can be applied to business as well. For example, the U.S. is able to produce a large number of people capable of planning and directing large international organizations. However, Britain, the Netherlands, France, etc., must form a partnership to direct large international organizations such as Shell Oil, Lever Brothers, Airbus Industries, etc.

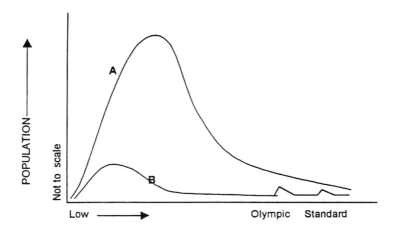

Figure 10.1   Athletic Ability. Reprinted from Jacques, E., A General Theory of Bureaucracy, 4th Ed. Heinemann Publishing Oxford. Ltd., England, 1986. With permission.

Since smaller groups are samples of the population of larger companies, it seems reasonable to assume that smaller groups within larger companies would have the same potential distribution of capabilities within a selected range, for example, creative and analytical people with outstanding abilities. However, the smaller the group the more difficult it may be to get a uniform sample with all the required capabilities. For this reason, one must be careful in forming a small group of say five people. They must be selected based on the information they may have rather than on their own capabilities.

Creative-Action Distribution

This brings up another important concept. If all the people in a large group are analyzed, it will be found that they fall into a grouping as shown in Figure 10.2.[3] It has been found that in any group about 75% of the people tend to let things pass them by and do not make a conscious effort to produce new ideas or to take any action to accomplish a task. However, about 12% of the people are action oriented and can break through the barriers to action. They don't seem to have creative ability, but they sure can make things happen. Give them an idea and a goal and you can't see them for the dust. No obstacle is too great; there is always a way around it. In today's society many of these people wind up in the news.

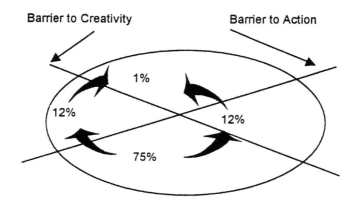

Figure 10.2   Creative Distribution

On the other hand, about 12% of the group are creatively oriented and have the ability to generate ideas, but they do very little with them. They may talk a good game, but they don't play. Many historians believe that Leonardo Da Vinci was such a person. Some people believe that he was one of the most creative individuals whoever lived but that he accomplished very little. He was so busy coming up with new ideas that he lost interest in those he had started and let them fall by the wayside as he jumped to the next idea.

It is believed that only about one percent of the people can break through both the barriers to creativity and action to make their ideas happen.

How can we find these idea-action people? We must build a team. The individual will not change overnight, but if we put the right mix together and build a team, the team will provide the effect. So, in any team, although knowledge is the prime consideration there is always room to add an apparently creative person or one who is constantly coming up with ideas or raising questions. In fact in some cases, the person you may want is one who is often thought of as disruptive because they always want to change things or do them in a different way. This is the time to make positive use of his constructive discontent.

So, we must work to build a team, not just a group of people but a group so involved that their brains are linked together to form a synergistic unit — five people working as one whose brains seem to be linked because the group has begun to understand each other's thinking. They begin to support one another for the overall benefit of the group.

We can see how difficult this is when we go back and look at athletics again. A football team may be a champion one year but not the next. A basketball team may win the championship for two years then disintegrate or keep winning but not become a champion. What is the cause? The key element is knowledge and trust. They must know the rules and trust each other to do their share of the job and support each member.

Team Organization

We spend many millions of dollars every day teaching people to work together as teams. However, this is mostly in the athletic business where team development has become a science. The first step is selecting the proper specialists with the knowledge and skills to perform the required task. Casey Stengel, the Hall of Fame New York Yankee manager, said, "It's easy to get the players. Gettin' 'em to play together; that's the trick."

So it is in industry; getting the players with the required knowledge and skills is the first step, but a group of people is not a team no matter how knowledgeable they may be. They must be trained and have a system to be able to work effectively as a team. Many companies now hire specialists to train their people to work as teams and most large companies have groups to train teams; some very large companies now have teams to train teams to train teams. At long last it is finally being recognized that just any group of people or a committee is not a team.

A committee is not a team nor is just any group of people a team. It takes three elements to create a successful team.

1. A significant project. Everyone must believe the project is worth the effort and that it needs to be improved.
2. Knowledge required to solve the problem. The necessary knowledge to solve any aspect of the problem must either be part of the group or available through consultants.
3. Commitment. The group members must make a commitment to solve the problem. Until there is commitment, there is chaos.

It is generally agreed by most authorities that a person's knowledge and abilities are limited to some degree. Whether the company sells insurance or builds automobiles, it is usually necessary to involve people of different disciplines to obtain the overall knowledge to solve problems. It is no longer acceptable for the engineer to design a product and "throw the design over the wall" to the manufacturing and purchasing people to obtain the necessary material and build the product.

For example, in a typical company there are design engineers and manufacturing engineers and these people may be further specialized by requirements such as plastics, steel, and aluminum or processes such as stamping, die casting, specific types of plastic molding, etc. There are also purchasing specialists, financial specialists in taxes, budgetary operations, and forecasting. Then there are marketing, advertising, product planners, and a host of other specialties. Although these people may be thoroughly knowledgeable in their area, they probably have only limited knowledge of the overall operation. In fact, there probably is no one in the company who has overall knowledge of the entire operation. To compensate for this specialization, we must make use of a team approach to problem solving.

There are innumerable methods designed to teach how to select and organize an effective team. Frequently, these instructions include profile tests to determine the best potential team participants from a personality standpoint. Unfortunately, you rarely get an opportunity to select people based on these results and in addition, those people with the best personalities for team participation may not have the necessary knowledge to solve the problem. Therefore, I have developed a profile based on knowledge, information, and responsibility as shown in Figure 10.3, Team Set-up.

Figure 10.3 illustrates typical team make-up for a product design, manufacturing analysis, or an administrative project. As already noted, every effort should be made to keep the team number to five persons. This may not always be possible but can frequently be accomplished by the selection of people who have dual experience. For example, in the manufacturing project the industrial and process engineer might be the same person, or the industrial engineer's expertise might suffice for the production foreman. However, when two different companies must be represented on a team, as when working with suppliers, it is frequently necessary to add an additional one or two persons.

The first person I believe must participate in a project is the one responsible for the project. For example, if the project involves a product design, the responsible design engineer must participate. If the project is a manufacturing process, the person responsible for the process must participate, and if the project is an administrative project, the person responsible for the system must participate. If these people cannot participate, then do not try to conduct the project. It is important to have the responsible person participate, not only because that person probably has more knowledge of the subject than any other, but also because that person will be required to approve recommended changes and participate in implementation.

There is a large group of people who do not agree with this approach to team construction. They believe that none of the team members should have been involved in any way with the original design. They believe the team should take an uninhibited view of the design. However, I believe

A suggested typical workshop team should include the following:

PRODUCT DESIGN

1. Product design engineer responsible for the product
2. Manufacturing engineer responsible for fabricating or assembling the product
3. Product engineer not directly involved with the product
4. Cost estimator or cost knowledgeable person
5. Last team member should be selected from the following group depending upon necessity and ability to contribute to the group.
   a. Quality specialist
   b. Materials specialist
   c. Industrial engineer
   d. Finance specialist
   e. Marketing-sales specialist
   f. Other - Often a spark plug

MANUFACTURING ANALYSIS

1. Process or methods engineer responsible or the project
2. Industrial engineer
3. Plant facilities engineer
4. Production foreman or supervisor
5. Finance representative
6. Last team member
   a. Product engineer
   b. Quality control specialist
   c. Material control representative
   d. Personal/ labor relations specialist
   e. Production worker
   f. Other - Sparkplug

ORGANIZATION OR ADMINISTRATIVE ANALYSIS

1. Person responsible for implementing recommendations
2. Person from major area that will apply the system
3. Person from second area that will apply the system.
4. Person knowledgeable with the system but not directly involved.
5. Personal, labor relations, finance or other area that may be involved
   Although it is preferred to limit the number of people to five participants persons from any area that interfaces with the system under study should be considered as a potential participant. Other areas to be considered may be purchasing, and manufacturing.

Figure 10.3 Typical Team Make-Up

the method used to review the project is designed to break down any mental constraints the original designer or other responsible person may have and assures a more thorough analysis of the project. In addition, I have often heard people who follow the "outside look" approach state that, when the original designer participated, implementation progressed more quickly and smoothly.

My primary reason for believing that the responsible person must participate is based on the process and on hundreds of successful projects.

The responsible person does not always freely agree to participate in the initial phase of the program because there is frequently the feeling that some of the other members of the group couldn't possibly contribute anything worthwhile to the project. This is where the system and the skill of the program leader become important. To achieve a dynamic, successful team, it is necessary for group members to participate in all phases of the deliberations.

There are also those who believe that they are rewarded for their individual performance and refuse to cooperate in a teamwork situation. They believe that they do not owe anyone allegiance for their knowledge but themselves. However, Albert Einstein is quoted as saying, "A hundred times every day I remind myself that my inner and outer life depends upon the labors of other men, living and dead and that I must exert myself in order to give in the measure as I have received and am still receiving."

I have discussed team participation and size with people around the world and found general agreement that the best size task force for solving problems is five participants. They may be supplemented by additional people with specialized knowledge that may be required during the course of the study, but there should be no requirement to call on them unless the task force team deems it necessary. However, every effort must be made to obtain people with the required knowledge and responsibility. The team members should have the distinct feeling that they are working to solve their problem.

The basic criteria for a well-organized team should be that there is someone on the team who knows something about whatever may come up during the deliberations or who knows where to get the required information.

In today's business climate in the automobile industry, every effort is being made to transfer as much responsibility as practical to suppliers. It is felt, among other things, that they are the specialists and the manufacturer–customer should make as much use of their expertise as possible. This is good thinking. However, it does frequently complicate the team construction by making it necessary to increase team size to satisfy all aspects of the project. Even though this may be the case, every effort should be made to maintain the magic number of five if at all possible and to never increase the number of team participants to greater than seven.

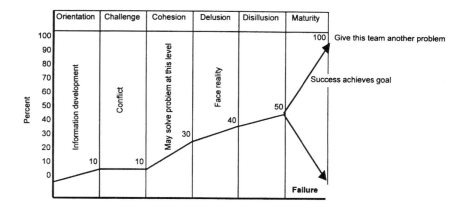

Figure 10.4   Typical Stages in Group Development

If discussions with the manufacturer–customer and supplier indicate the project requires more than seven people, it may be that the project should be redefined and two teams assigned to two projects. It might also be preferable to assign the same basic team, with necessary replacements, for the second project team and conduct the second project at a later date. This second project should begin within a few weeks after completion of the first while the previous project knowledge is still clear in their minds.

Once the workshop group has been selected and assigned to a task it will still take some time for them to come together as a team. Initially some members of the group may not know other members, and there will be an air of reluctance to expose themselves to the other members in the group. This reluctance was referred to in Human Relations, Figure 9.7, the Johari Window.[4] Since each person was selected to participate because they have some particular knowledge necessary to the group, this reluctance will break down as the program develops. This is a normal situation.

Team Development

There are four distinctive steps in group development as illustrated in Figure 10.4, Typical Stages in Group Development.

The First Step — Polite stage.
This is when the group gets acquainted and feels each other out.

The Second stage — Answering the question, why are we here?
   The group defines goals and objectives.
Third stage — Bid for power
   This is when each member strives for influence and control. The
   individual members are competing to determine the leader.
Fourth Stage — Unity.
   The group has developed mutual acceptance and has begun to act
   as a cohesive group. In addition, they have developed *esprit de
   corps*.

As the group develops into a cohesive team they follow the path
identified in Figure 10.4. In working to solve any problem, there are always
stages when the group feels they have the best solution to the problem.
Then, unfortunately, they find the idea has a serious flaw and simply will
not work. This can lead to frustration and disillusionment and ultimately
failure. However, if the group understands that this is the very nature of
the process and that the group did not fail, only that the idea did not
work, they can go on to greater heights and ultimately solve the problem
with far better results than they ever expected possible. The next step is
to assign the group another project now that they have learned to work
together and feel they can do anything.

The first step in a project as well as team development is to assign
the group to review available material regarding the project. This will
require one or two people who are the most knowledgeable in the project
to lead the group in this initial stage. At the same time others in the group
will be developing questions based on the discussion. If you refer to the
Johari Window in Figure 9.7, Motivation, you will see that there will be
questions about the project that someone else knows the answer to but
you do not. There will also be questions that you know the answers to
but that others do not. In this early stage the objective is twofold; to learn
as much as possible about what is known about the project and to
distribute this information to everybody in the group.

A typical list of information that should be available to the group is
shown in Figure 10.5, Basic Project Information. The list shown is for a
product and may vary somewhat for an administrative project. Although
all of the information available to the group may not be used, it is better
to have it available and not use it than not have required information
when it is needed and have to waste time looking for it during the course
of the project.

There are also cases when there is not much information available on
the project. This may be the case in public service or administrative
projects where the project is not clearly defined. However, no matter what

Required Project Information

1.   Assembly and parts drawings

2.   Technical information (Forces, stress allowances, etc.)

3.   Quantity requirements, annual usage

4.   Sample of assembly, breakdown showing parts

5.   Cost data - Material - labor - variable burden breakdown for each
     part of the assembly. If variable burden is not available,
     provide fringe benefits as a percentage of labor.

6.   Tooling cost

7.   Manufacturing data - planning sheets (sequence of operations)

8.   Specifications (Company standards and government requirements)

9.   Market surveys (Focus group information)

10. Operating service history (warranty data, etc.)

11. Special features or requirements

12. Future plans and objectives

13. Competitive situation (Benchmarking data)

14. Other potentially useful information

Figure 10.5   Required Project Information

the project may be, it is important to collect as much information as possible before the first group meeting. In some cases, it was found shortly after initial deliberations had begun that the project was not what it seemed. At this stage the project was redefined, recommendations were made to reconstitute the team to satisfy the newly redefined project, and required information was listed for collection.

Nontask Team Issues

An excellent example of an effective team in the athletic field would be a basketball team. A basketball team is made up of five players and the coach. Each player is a specialist. There are the center, the two guards, the two forwards, and the coach. The coach develops the strategy, and the players execute the strategy to achieve a common goal; win the game. Under normal conditions this isn't an instant achievement. It takes time for the coach to learn the player's abilities and shortcomings, and it takes time for the players to not only learn to trust the coach, but they must also learn to trust each other.

Although teams can be extremely effective, they do not just happen or evolve because people work together. They are very complex organizations that continually require work and nurturing. All members must realize that whether they succeed or not depends not only on their technical skills, but also their interpersonal relationships with other participants. Consequently, every member must gain understanding of the importance of these interpersonal relationships and how these relationships can affect the success of the team.

Teams, unlike committees, have certain inherent properties that make groups of individuals excel. Bill Russell of the Boston Celtics has been quoted as saying that, "We were a team of specialists, and like a team of specialists in any field, our performance depended both on individual excellence and how well we worked together. None of us had to strain to understand that we had to complement each other's specialties; it was simply a fact, and we tried to figure out ways to make our combinations more effective."

When the team members work well together, the team's abilities become greater than the sum of all its members and therefore, they are able to accomplish more than any one person can. Each individual brings his or her own expertise, but rarely does one person have all the information or experience to completely know every aspect of a process. By bringing people together from all levels and areas within the organization, the group will gain a well-rounded knowledge of the task When people pool their resources there can be an increase in productivity and quality that produces a synergistic effect. Through the use of teams, organizations are better able to identify multiple processes or problems that need improving. Teams help because teams draw out people's skills.

The athletic business is not the only place to find teams. A symphony orchestra is also a team with the very same problems and relationship requirements as an athletic team. Although larger than we like to use for problem solving, a top flight orchestra usually runs in the order of 100 members. A recent article noted that the New York Metropolitan Opera orchestra was, in the past, considered to be only adequate; the primary concern was for the voices. However, the new conductor, James Levine,[5] turned "an ugly duckling into a swan" according to the critics. How did he do it? He concentrated on technique and details. Over and over again he concentrated on perfection. To a large degree it was a matter of improving attitude. Good enough wasn't good enough for him, and he thought it wasn't good enough for the talented musicians in the ensemble. The result is a superior orchestra that blends with the opera to make for a truly outstanding experience.

How to Become a Team

Practice makes perfect. Even the smoothest running teams have their ups and downs. It's all part of team development. Successful teams are like a well-greased machine, but they need constant attention. Like any relationship between people, team membership takes work. Each member knows that the team's success depends both on individual performance and their ability to work effectively with others.

When a person joins a team, they feel excitement, apprehension, and confusion of loyalties. If these emotions are not addressed, the team will waste time and will become inept. According to Peter Scholtes there are three interpersonal issues common to all teams. The first one is personal identity in the team. Each member worries about fitting into the group. The second issue is relationships between team members. Participants worry about whether the different members will get along, especially since they come from different levels and different areas within the organization. The last issue is identity with the organization. Up to this point a member's identity has been with a department. How will this new membership affect future relations, duties, and responsibilities in their department or in the organization in general? Once these feelings and emotions are out in the open and addressed, the team can become more focused on the task at hand. Like anything, the first step in solving a problem is knowing that it is there.

References

1. Lindbergh, C. E., *Spirit of St. Louis,* Scribners, New York, NY, 1953.
2. From Jacques, E., *A General Theory of Bureaucracy,* 4th Ed., Heinemann Publishing Oxford England, 1986.
3. Jonelis, J. A., Personal discussion, 1988.
4. Picraux, F. J., Chrysler Corporation Value Engineering Workshops, Personal communication, 1972 to 1975.
5. Ames, K., The making of an ensemble, *Newsweek,* Sept. 30, 52, 1991.

Chapter 11

Creativity: The Innate Drive To Change

Introduction

What did Da Vinci, Galileo, and Einstein have in common? The ability to develop new ideas and to see things in a way that nobody had ever seen them before. They were able to look at one thing and see another, which is a basic ingredient for creativity. Because of this talent, they are seen as gifted and often referred to as geniuses, which means someone with extraordinary mental superiority. However, this creative talent is not given to just a select few. We all have it. The only difference between us and these three men is that they harnessed and developed their creativity into channels that brought them recognition. We have let our creative ability go dormant.

What would we be like if we were able to redevelop these creative powers? Would we be able to perform like a genius? This is a provocative question. What is the ingredient that makes it possible for some people to break the barriers to visualization, to look at one thing and see new things; to work as a total being with the conscious and subconscious working together? It is, of course, the way we use our brain.

In spite of the fact that the future depends on creative solutions to problems and creative applications to develop new products and systems, we put very little effort into increasing creative abilities.

Although we hear much about creativity, just what does the term mean? The dictionary says to create is to bring into existence something new. It also says creation is the art or practice of making, inventing, devising, fashioning, or producing an original work. Creativity, therefore, is having the power to be creative and develop new ideas. Although this may be the dictionary definition, there are dozens of other definitions that range from the ability to develop a new viewpoint to applied imagination. However it is defined, it is agreed that a creative act results in a product that is new or original or different from one that preceded it. It is the process by which original patterns are formed and expressed. The key word is original, the idea does not need to be new because there are times when people working in widely separate locations simultaneously develop the same idea.

Most authorities agree that we are all born creative. However, much of our creativity diminishes as we grow older. This is partly due to self-imposed constraints we apply to ourselves in order to be accepted as one of the group and fit into society in general. The Japanese have a saying that clearly expresses the problem, "The nail that sticks out must be driven back in." Creative people "stick out," so to help regain some of our creativeness, we must overcome the personal and environmental factors that stifle it.

Many of us also fail to be creative because Western education has tended to make us focus on analysis rather than on creativity. It has been clearly shown that the average person reaches his or her creative peak at about seven years of age, just after entering school and being taught the importance of conforming. It is, of course, necessary to conform to certain standards of social behavior, but think of what a Van Gogh would look like if all his paintings conformed to blue skies and green grass.

It is generally considered that there are three types of reasoning: deductive, inductive, and judicial.

1. *Deductive reasoning* is based on drawing conclusions based on facts and available information. It is basically analytical and is the primary method taught in school. The chief concern is with finding the one correct answer to a problem.
2. *Inductive reasoning* is used when there is insufficient available information. Several concepts are developed and the thinker adds ideas of his own imagination. This is the basis for creative thinking.
3. *Judicial reasoning* is based on critical thinking. It is basically analytical.

Although the importance of our analytical skills should not be under-estimated, they are only one aspect of our thinking process and are sometimes over-used as well as misused. Only a few of us have learned how to use our inborn intuitive and creative capabilities. Many of us suppress our creative impulses because we want to fit into society; still others want to be reasonable and don't want to fail. However, it is important to note that making greater use of our latent creative powers can provide better balance in our thinking and lead to improved performance.

Left and Right Brain Theory

The human brain is divided into two hemispheres. The hemispheres are symmetrical in shape but not function. The left hemisphere controls the right side of the body, and the right hemisphere controls the left side. This distinction has been recognized for years. However, the role the two play in the mental process is something new. Tests such as Positron Emission Tomography (PET) are giving us a better insight into how the brain functions and the different roles the two hemispheres play.

With the PET scan patients are injected with radioactive tagged substances which emit gamma rays. A tagged form of glucose, the brain's fuel, is often used because glucose uptake and use are higher in active areas of the brain. A computer detects and records the gamma rays and maps the brain's biochemical activity. PET scans show not only how the brain is affected by disease, but also which hemisphere is active when normal people are listening to music, moving their arms and legs, or just looking at their surroundings.

Researchers have now found that, in most people, the left half of the brain is used to analyze and to solve problems in a step-by-step manner. It also controls our verbal skills. The right hemisphere is the creative side and is devoted to problem finding, esthetic judgment, and intuition. It tends to process information that cannot be put into words and thinks in terms of whole patterns, even when only partial information is available.

The findings regarding the function of the brain cannot be minimized. Most of the knowledge we obtain is the result of the activity of the left side of the brain, which translates creative ideas into words. Our creative ideas evolve in the right side of the brain and the left translates them into communication. The lesson, then, is that both brain hemispheres need to work in harmony if we are to produce creative solutions to our problems.

Much of Western education has focused on logic and reasoning, the function of the left half of the brain. Studies show that teachers only give students two to three seconds to answer a question and that isn't long enough to think the question through in the right hemisphere. Many feel that more emphasis should be placed on the right side of the brain. It is believed that if the creative side of the brain was strengthened, the effect could be fantastic. Dr. Cecil Reynolds[1] at the University of Nebraska, states, "The total power of the two hemispheres working together is much greater than the sum of the parts working alone."

Up to the third grade, education emphasizes both sides of the brain. However, the emphasis switches to the left after that. To gain the advantage of using both hemispheres, Dr. Reynolds urges parents and educators to go back to basic skills and a balanced approach to education. This includes such activities as art, music, and physical education. Many people see these as frills, but Dr. Reynolds sees them as essential.

Dr. Tadanobu Tsunoda[2] of Tokyo Medical Dental University suggests that there is evidence that the differences in cultures between East and West can be traced to the different ways of using the brain. It is also known that engineers are left brain or logic and reason oriented and architects are right brain oriented or devoted to creativity, aesthetic judgment, problem finding, and intuition. Therefore, when working together, although their backgrounds may complement each other, they look at things differently and may have difficulty understanding each other's viewpoints.

Invention and Discovery

Invention and discovery are both by-products of creativity. Creativity occurs when a person sees something new in the environment that surrounds him. These insights can result in an invention, the creation of something new or original such as Watt and the steam engine, Bell and the telephone, or Edison and the electric light. Creativity creates new and original results; it reveals a discovery. These new ideas may be the result of inspiration or serendipity. Edison claimed his inventions were the result of 10% inspiration and 90% perspiration. Although Charles Goodyear had been working on a way to improve rubber for many years, his discovery of vulcanizing was serendipitous as the result of an accident while working on a batch of rubber on the kitchen stove. Goodyear could have said, "Another batch ruined!" But he didn't. He immediately recognized that heat was the solution to his problem. The heat "vulcanized" the rubber.

To further illustrate the above point John W. Haefele[3] in *Creativity and Innovation* uses a simple mathematical equation to describe the creative process. Equations are an excellent way to consolidate information and quickly show a specific idea. They are flexible enough to illustrate any concept, even creativity.

In the equation A and B represent two independent concepts. If they are united they will form a new concept, C. Our formula for invention, therefore, will be A+ B $\rightarrow$ C! The arrow means to yield or produce and the exclamation mark following the C indicates a new and often surprising finding. Therefore, C! is the result of combining A and B.

Obviously, the more information or concepts available the more potential solutions may result. This combination of concepts to form something new is called invention. However, it is also possible to start with C and discover A and B, which may be new concepts. This is called discovery. So we have two basic processes, invention and discovery. This leads to a slightly different equation, A+ B $\leftrightarrow$ C!, indicating the two-way possibility.

The information for the A and B input comes from a lifetime of awareness of your surroundings. In fact, some authorities believe lifetime experiences and learning are stored somewhere in the brain. The trick is to retrieve the information when it's needed. Some proof of this theory is found in a little exercise you can try, and that is to recall the name of your third-grade teacher. It is surprising, when I ask this question of a group of people in a seminar or workshop, how many people can recall the name almost immediately. In most cases, the majority can recall the name within a few minutes.

A good number of years ago the research laboratories of a major oil company decided to try to determine just what the ingredients were that made a creative person. They had more than a thousand PhDs in the laboratory, which was a pretty good sample. They asked what was it that made it possible for some of these people to produce hundreds of patentable ideas, while others with exactly the same qualifications produced only a few.

The background, education, hobbies, and personal habits of each person were reviewed and no conclusive result was ever determined. However, there was one characteristic that appeared to offer a clue. Those people who held the most patents seemed to read more and a wider variety of material than the others. Everyone read scientific journals and material relative to their specialty, but in addition, those with the largest number of patents read comics, novels, biographies, newspapers, magazines of all types, and anything else they could get their hands on.

If we consider the simple formula above this would seem reasonable. The prolific readers were soaking up more varied information to use in their work. It is, therefore, likely that their minds should be open to a wider variety of experiences.

This would seem to indicate that to improve our creative ability we should practice to improve our awareness, not only of things that affect our job but of everything around us. It has been said that in times past the native Indian was constantly aware of the things around him. As he traveled through the wilderness he was aware of animals that had passed recently, the direction they were traveling, how many there were, and many other details of value to the hunter. We know the police officer is trained to be sensitive to certain things as he makes his daily or nightly tour of his assigned area. In many, if not most cases we do not see what either the Indian or the police officer sees. Why not? What do we have to do to increase our awareness? The answer is simple — practice.

Creative Characteristics

The Creative Person

People who have cultivated and developed their creativity can come from vastly different backgrounds and environments. Despite the many differences among creative people, researchers have found that truly creative people share some common characteristics.

Awareness — Creative people have inherited sensitivity to such things as mathematics, art, music, and literature. In many case these people come from families that have passed this sensitivity from generation to generation. In addition, they tend to have sensitivity for almost everything around them. They seek out problems rather than just find answers for those that exist.

Childhood Training — Most of our attitudes are developed before the age of seven. Creative people are brought up in an atmosphere that encourages inquisitiveness and individual thinking. They are not afraid to ask questions and may not have become regimented in their ideas.

Education — Much like early training, a premium is placed on curiosity rather than conformity. Learning to ask questions vs. finding answers is emphasized. Rather than a technical education with emphasis on the right answer, a liberal education based on finding and answering the right questions seems to be preferable.

Tolerance for Chaos — Creative persons have the ability to make order out of disorder. They can arrange multiple experiences into meaningful patterns. However, when faced with a large amount of data, a creative person is likely to come up with a widely different solution than the rest of the group.

Personal Courage — The creative person is not afraid of failure or of being laughed at. Criticism can even be a stimulus. This may be the result of early training that fosters inquisitiveness and self confidence.

Sustained Curiosity — It might be said that a creative person is some-one who has never grown up. He or she keeps asking questions and feels excitement when facing the unknown.

Dedication — Creative people are determined to find the answers to problems and to express all their beliefs. No obstacle can keep them from reaching their goal. For creative individuals, time is relative. Solving a problem could take a day or decades. Morning, noon, and night are the same to a creative person.

It can even be carried into the subconsciousness of sleep. They are self-motivated and have an intense drive to complete the project.

Independence — Perhaps one of the most important characteristics of the creative person is independence. There is never a fear of being different or being ridiculed. What is important to creative people is what they think of themselves, not what others think of them.

In addition to these eight characteristics, researchers also suggest that creative people have fluency, originality, flexibility, and most important, an unguarded self conscious, which means letting one's subconscious thoughts surface to the conscious level. It is in the subconscious where the truly creative ideas are formed.

The Creative Process

Seven Steps To Creativity

Our formula, $A + B \leftrightarrow C$! is one way of describing the creative process. Although some people say there are no rules for creativity, there does seem to be a process that the system appears to follow to complete a creative act. The seven steps are: desire, preparation, manipulation, incubation, intimation, illumination, and verification.

Desire is the first step in any creative act — the need to create something new. The reasons for this desire can vary. A person may create because he needs to know or must solve a problem. Whatever the reason, he or she must be motivated by some inner drive. Without it, creativity cannot get started.

Preparation is the next step. This is when we develop the A and B in our formula. It requires gathering as much information as possible, and for maximum benefit requires the application of both our analytical and creative skills. The information gathering may require research, experimentation, or just the personal experiences of being alert to our surroundings over a lifetime.

Manipulation is when a creative person tries to use the information available by reworking it to develop a solution to a problem or making something useful that has never existed before. He or she tries to alter his or her outlook by shuffling data around or seeking analogies and metaphors that provide new insights and produce new ideas.

Incubation is the step that involves the sleeping solution. Solutions rarely appear immediately. Often the problem is dropped, and the person moves to another project. However, the project is not forgotten. The person lets the subconscious play with the problem. In the incubation step, manipulation of information still occurs, but at the subconscious level.

Intimation means that we have faith that a solution will be developed. It takes place in the subconscious where most creative ideas are developed. For ideas to be activated, they must make their way to the conscious level. This takes faith and an understanding of the creative process.

Illumination is the most exciting of the seven steps. This is where the solution to the problem is realized. It is the "C!" in our creative formula. The "AH HA" phrase is often illustrated by a light bulb. The light has now been turned on. We see the light.

Verification is the last step. The new solution, idea, or creation must now be tested to see if it accomplishes the required purpose. If it fails to work, the person may have to start over or at least determine what has been learned in the course of the failure. However, if the project does succeed, the person will feel accomplishment and success. Verification may encompass many failures, but faith in the idea drives the creative person forward to a successful conclusion.

These seven steps are the road to creativity. Although they are listed as separate and distinct steps, they are often performed subconsciously; in addition, each step is not always required for every creative act. Sometimes a person will get lucky and find a solution before going through all the steps. This will happen more often in discovery than in invention. For the most part, however, the sequence is usually followed.

A homely example of how the system works can be seen in the simple process of making coffee. Although this is not a creative act because it is repetitive, it does illustrate the process.

Preparation — Assemble the necessary materials — coffee maker, coffee, and power.
Manipulation — Organize the components in the proper amounts.
Incubation — The sleeping solution. Wait for the process to progress.
Intimation — Have faith that the process will produce the desired result.
Illumination — The coffee is finished — C!.
Verification — The coffee is ready for use. Test and apply.

Although the process seems simple it doesn't always work that way. During a TV program on creativity, Bill Moyers was interviewing Harold Black,[4] a research scientist who holds over 300 patents. Mr. Black was describing the morning he came up with the solution to improving the voice quality in amplified long distance telephone circuits. Mr. Black was going to work at the Bell Telephone Labs in New York City one morning in 1927. He was describing how relaxed he was when he was on the ferry looking out at the water and the sun glinting off the windows of the buildings in the city. Just as the Statue of Liberty came into view he said, "Bam, the idea hit me." He did not have any paper to write on, so he grabbed a fellow passenger's *New York Times* and drew a circuit diagram and several equations on the front page and rushed to the office where he had the idea witnessed. He realized that he could build a distortion-free amplifier by tapping off a small part of the amplified signal and subtracting it from the input signal.

In time, his idea developed into the theoretical understanding of a new and major engineering field of feedback-control systems. However, it took 12 years to obtain the initial patent because so many learned people around the world said you couldn't do that and it wouldn't work. All the while, Mr. Black said they had over a dozen machines working perfectly in the lab.

So the point of the story is that Mr. Black was in the intimation stage when his idea came to him. He was thoroughly familiar with all facets of the problem and had taken his information, twisted it, probed it, and could do no more. However, his subconscious was working on the problem under his belief that something could be done. It is reported that Einstein once said that the older you get the more you come to recognize that you cannot impose your will upon nature. However, one day while you are sitting under a tree eating an apple, the idea comes up to you and says here I am.

However, the idea doesn't complete the cycle; the idea must be proven and implemented in some way. This is the verification phase, and the idea person must have confidence and dedication to overcome the judicial thinking that will try to prove that the idea will not work. In Mr. Black's case this took 12 years, but it was worth the struggle.

Blocks to Creativity

It is impossible for us to be the best of everything because of obstacles we encounter. For example, we might be able to read music, but with poor coordination we will not be able to play an instrument well. To improve our performance we must identify our problem, which is coordination, and work to improve it.

An analogy can be made between the above example and creative thinking. We all have conceptual blocks or mental obstacles which make it difficult for us to correctly perceive a problem or to develop creative solutions. The quantity and intensity of these blocks varies from person to person. If we wish to improve our creative problem-solving techniques, we must first start by identifying our mental obstacles.

We all have mental constraints of some sort; many that we have applied over a period of time to fit into society. If you don't believe you have any mental blocks, try the simple test below. Read the short message and count the number of F's.

FINISHED FILES ARE THE RE-

SULT OF YEARS OF SCIENTIF-

IC STUDY COMBINED WITH THE

EXPERIENCE OF MANY YEARS.

If you counted six you are correct. However, many people can only find three or four, even after they have been told there are six. This is a mental block that is self imposed, in that we often do not consider what may be thought of as unimportant words such as OF.

James L. Adams[5] identifies a number of blocks that fit into the six categories listed below.

Perceptual	Emotional
Cultural	Intellectual
Environmental	Expressive

Perceptual Blocks prevent a problem solver from clearly perceiving either the problem itself or the information necessary to solve it. We have found that in their anxiety to solve a problem many people never really identify the cause and only work on the symptom. In addition, they restrict the scope of the problem by limiting it to familiar systems and concepts. In other words, they frequently try to make the problem one they know how to solve. There is also a tendency by many to overlook others' viewpoints. In many cases people see what they expect to see and ignore anything else. Their solutions are typically based on brain logic with no time for right brain creative solutions with strange unfamiliar inputs.

Cultural Blocks are behavior standards that cause us to think and act in certain ways. In many cases these taboos have been set up to protect society from danger. Many have helped to prevent the spread of disease and others to improve social contact by eliminating chaos. Whatever the reason, they can limit the range of thinking and possible solutions to problems. If a solution might go against a taboo, it may not even be discussed, much less considered.

Although these taboos may lose their impact over time, a very similar block can be tradition. In industry, tradition is a strong cultural block. It may be that a company has developed an outstanding capability in a process. As a result they try to use that process as much as possible and in many cases their investment causes them to neglect other new and improved methods.

Environmental Blocks are mainly related to the workplace or wherever you may be expected to do your thinking. In the office the telephone is a primary distraction, and now it is possible to add this distraction to your car to destroy one of the best areas for creative thinking we have available to us. Other distractions are office people, traffic, manufacturing noise, and vibration. Creativity needs a place to concentrate and it's up to the creative person to find it.

A lack of support in the form of no interest on the part of management for new or different ideas can also kill creative effort. It might be that the boss does not accept other people's ideas, or at least ideas with which he is not familiar, or that there just isn't enough time to investigate new ideas.

Emotional Blocks are headed by fear. The inability to do something for fear of embarrassment is an emotional block that frequently prevents people from conceptualization or at least publicizing their ideas. This is a basic psychological problem and one that can be very difficult to break. However, mistakes and failures are a part of creativity. When Edison said that his work was 10% inspiration and 90% perspiration, he was saying he tried and tried until he found something that worked.

Jack V. Matson at the Leonhard Center for Innovation and Enhancement of Engineering Education at the University of Pennsylvania says that risking and accepting failure is part of the creative process. He says that if you want to be successful, the key is not to avoid risk but to manage your failures.

There are also those people who have become authorities in their field. They know all the possible ways to do something and become very judgmental of ideas. If it does not fit their pattern, it will not work. When faced with a new idea their first reaction is to tell you why it will not work. Wait them out and when they have run out of negative reasons, ask what has to be done to make it work. In many cases this makes them a partner with you and often leads to success. We must learn the habit of finding at least one good thing about a new idea.

Intellectual and Expressive Blocks cover two important factors. The first is trying to solve a problem with the wrong method. An example is applying a complex mathematical process to a problem that could be solved with a simple diagram. The second is one of the worst blocks that a problem solver can have — inadequate or incorrect information. There are times when the misleading data can come from the highest sources, which may tend to provide credibility to the information. It is the problem solver's responsibility to verify all information to be sure the accuracy is adequate for the purpose and that it is up to date. If the information we have is incorrect, the outcome will most likely also be incorrect.

Improving our Creativity

Many of the obstacles discussed above can be, at least partially, overcome with practice. Some might have to be accepted and tolerated. The most important ingredient to overcoming blocks, however, is to keep a positive attitude. You must believe that you can and will overcome them. You must constantly work at eliminating obstacles. This will require time and patience. In *Conceptual Blockbusting,* James L. Adams lists the blocks to creativity and gives simple puzzles and diagrams to help you identify your blocks. He also lists things that you can do to break them down.[5]

The use of our subconscious can be one of our best tools for being creative. When Frederick Kekule,[6] a famous German chemist, was unable to make any progress in his research, he began to daydream as he stared at the fire in the fireplace. It was at this point he discovered the structure of the benzene ring.

Listening to our subconscious and having faith in the ideas that develop from it requires self-confidence, Maslow[7] suggests gaining self-knowledge. The result will be improved confidence and creativity. He bases this on the fact that things are less threatening when people are convinced of a positive outcome. Postponement of criticism or judgment, therefore, is another important ingredient in unleashing the subconscious.

Fluency and flexibility are two interrelated methods for overcoming obstacles. To practice these techniques, the simple task of list making can be done. List making tends to use the compulsive side of a person, resulting in making us effective conceptualizers. List making also makes us see a concept in more than one way. Fluency and flexibility of thinking takes a conscious effort, and list making can make us more effective at both. It gives us more insight into the possible function of an object, which in turn, is an advantage when developing ideas. Do not look for the best way; look for 100 ways.

Our conscious awareness of things that surround us can also play a role in creativity. By making a conscious effort to challenge assumptions, asking why, looking for patterns, exchanging ideas, and letting the right side of the brain create, we can hurdle over obstacles. Ironically, it is only with conscious effort that we can learn to use our subconscious and the other tools that make us creative.

It has long been known that creative people do not work by the clock. When 5 o'clock comes around they may leave the office or lab, but they do not necessarily shut down the shop. They continue to think subconsciously about their work, problems, and objectives. In most cases this self motivated, subconscious effort is inherent in the individual; some are better at it than others.

History is strewn with examples of people who made use of this subconscious to solve problems or to come up with new, startling ideas, and they are still doing it today. Ken Perlin[8] recently won an Oscar for writing a computer program called Perlin Noise, which is important in the entertainment industry. He came up with his idea at 4 o'clock in the morning when pouring cream into his coffee. He noticed that the swirling coffee looked like marble. He took this idea and applied it to his computer and developed his problem solution.

In any case the idea is that relaxation in some form or other can aid in creativity. Some have solved their problem while drowsing, others while walking. Friedrich Nietzsche[9] has said, "All great thoughts are conceived while walking." Still others believe that great thoughts and solutions to problems can be conceived while sleeping.

Several years ago while driving I heard an interview on the radio with a woman who had kept a dream diary for over 30 years, starting when she was seven years old. This diary led to her becoming a psychologist specializing in dreams. She pointed out that there was a primitive tribe in the southeast Asian jungles that taught their children how to dream to solve problems, and she discussed some of the details of the training. These experiences are described by Patricia Garfield[10] in Creative Dreaming.

The subject triggered my curiosity and some time later while on a business trip I came across an article, *Mind Over Mattress* by William F. Fry, MD.[6] Dr. Fry pointed out that many people have trained themselves to deliberately solve problems and create ideas in their sleep. He cited Robert Yates, a Lockheed engineer with over 100 patents to his credit, who says he developed his concepts in his sleep and immediately upon awakening in the morning wrote them down.

Dr. Fry describes how he designed his garden fence by trying ideas in his sleep until he found one that he liked. He pointed out that the dream brain has been productive in scientific research as well as in literature. The problem is how do you develop this ability? He says it is an automatic, innate process, and there are some guiding principles to help make it work how and when you want it to. However, most of all you cannot force the issue.

To make the system work, he says gain as much information as is available on the subject, then wait several days to let the information sink in. This is the incubation stage of the process. On the evening you want the dream, review the information before you go to sleep. To increase the chance of success follow the specific recommendations below.

- No alcohol.
- Ensure uninterrupted sleep.
- Choose a night that has been free of dreams for several previous days.
- Assure moderately cool bed temperature.
- Avoid exciting experiences.

To recap the principle considerations:

- Don't try to force the issue.
- Avoid a stressful preliminary period.
- Relax. It may take several days.
- Test and experiment. It takes time and patience.

Although Dr. Fry's process is reasonable, everything I have read indicates that creative dreaming takes time. It takes time to learn the procedure. It takes time to learn how to relax. It also takes time to capture the idea.

A child should be able to learn the process and begin to develop successful results within a few days, but for an adult several weeks to a month is more likely. Some may never be able to develop the technique because they have too many mental blocks. So it isn't a matter of just following the procedure. It is a creative method and requires all of the preparation necessary for any creative act. Desire and motivation are primary components.

Stimulation of Creativity and Creative Techniques

Games can play an important part in activating creativity and the conceptualizing of ideas. Creative games fall into two basic groups. The first group includes games that force an individual to draw upon his knowledge and experiences, with creative talent acting as a catalyst. Among these are the Gordon Technique, Checklisting, Inversion, Morphological Analysis, Area thinking, Input–Output, Alphabetical listing, and Brainstorming. The second group includes games that involve activities designed to stimulate or teach us how to use our creative powers. These include various puzzles requiring us to draw lines, visualize squares, move coins, etc. A sheet of several of these games can be found in the Appendix.

Of the first group of games, we found Brainstorming to be extremely useful. It helps a group of people whose creativity needs a boost to come up with a great volume of ideas. These ideas are not answers; they must be screened and evaluated to become concepts on which firm recommendations can be built. This process generates an atmosphere that permits each person to freely depart from his logical and conforming mental control. It is in this process that you may hear yourself differently and perhaps feel uncomfortable, but everyone shares this feeling. If you can persevere and stretch your tolerance for some discomfort, you will realize the beginning of the potential you have for real creative productivity.

You will have plenty of help and cooperation in a brainstorming session, and it may demonstrate capacities you have not yet realized. However, in order to be successful, it is necessary to follow a specific set of ground rules. These are necessary to ensure the proper environment for idea generation and development.

A good tune-up to set the stage for a brainstorm session is to have the group try a simple exercise. Ask the group to name as many types of ships as they can think of and continue for several rounds. The group will start out naming the familiar warships, liners, tugs, etc. As it becomes more difficult to name different types of ships the thinking will become more intense and someone may say scholarship, then the next person penmanship, etc. This will demonstrate "hitch-hiking," building on someone else's idea. Not only does an exercise demonstrate several of the brainstorm techniques, but it puts everyone in a free thinking mood for the main session. Of course you can only use this specific exercise once with a group; however, you can build your own set of techniques.

The first requirement for a brainstorming session is to isolate the question and/or questions that will form the basis for the session. These questions must be clearly identified. Well-defined project functions are usually used as the question or the basis for questions. The question is then presented to the group, and the group tosses out ideas regarding the question. Any idea is acceptable and no discussion is allowed. In addition, it is important that as many of the ideas as possible be spontaneous. There should be no attempt to evaluate any response at this stage because it will kill the spontaneity. The intention is to come up with at least one new idea that has never been thought of before.

The following ground rules are necessary to ensure a successful brainstorming session.

Brainstorming[11] — Ground Rules for Success

1. No criticism allowed during session.
2. A peer group is usually best. Never have high-level management and their employees attend to avoid intimidation.
3. Quantity is desired. The more ideas the more likelihood of at least one outstanding item.
4. Six to 10 participants is best.
5. No publicity on the session after it is completed.
6. Combine ideas. Hitch-hike on other's ideas.
7. Wild ideas wanted. The first 90% of all ideas will be those that have come up before.
8. Record ideas. Paper is best.

It is best to record the ideas in writing. A tape recorder sometimes seems to intimidate the group. However, a computer can be very useful. A typical list may run over a hundred ideas and a computer can aid in tabulating ideas and in screening and transposing the final list.

Blast, Create, and Refine

A creative technique that has been proven to work successfully is Blast–Create–Refine. This technique not only works well but is an outstanding process for simplifying or reducing cost. The reason is that it starts with the basic requirement and adds only what is required to meet the objective.

In many cases people have been working to reduce the cost of a product. Usually this effort results in a 5 to 10% reduction and is achieved by reducing metal gage, changing material or some specification, or by eliminating some element of the product. Occasionally, it may be possible to reduce the cost by a maximum of 15%. However, this will probably require some major product change that will add an additional expense for tooling or other equipment. One of the results of this situation is that these percentages have come to be accepted as the best that can be done.

If we are looking for a larger percent reduction, like 50% or more, a completely different approach is required. The Blast, Create, and Refine technique is an effective system that incorporates function definition, creativity, and function analysis or the evaluation of ideas to find more effective ways to achieve an objective.

It is a fact that if very large cost reductions are to be made it is necessary to start at the bottom and look at the basic concept rather than trying to reduce the cost of an existing product. We have covered many of the reasons for this earlier in the text. Intense review of a product shows that it is, to a greater or lesser degree, the result of a series of decisions made during its evolution. Therefore, the search for better value requires that we ask, How can this chain of influence be stopped? How can we objectively look at a function?

The Blast, Create, and Refine technique makes possible creative solutions, first by eliminating details of the existing product and then by freeing the mind for creative activity. Second, it directs thinking toward basic considerations. Third, it provides a method to build on these basic considerations to develop a final product satisfying all necessary requirements.

An example will illustrate the three phases of the process.

1. *Phase 1, Blast* — This requires defining the basic function of the product or segment of the product under study. Next, blast that portion of the product out of the problem so it can be clearly identified. This means identify the basic function, the one that most clearly defines the requirement and offers an opportunity for many creative ideas.
2. *Phase 2, Create* — The second phase is to develop ideas to satisfy the basic function and select one that appears to best satisfy the basic requirement, ask a series of constructive questions. The questions are as follows:
   What is the lowest-cost way to satisfy the basic function?
   Will this idea satisfy the requirement?
   If the answer is no, ask, "What has to be done to make the idea work or sell?"
   Alternatives are developed and the same questions asked of each alternative until a satisfactory solution is achieved. Make no attempt to evaluate solutions at this time.
3. *Phase 3, Refine* — The refine phase of the technique is to take what appear to be the best alternatives and refine them to a final recommendation. This will involve preliminary cost estimates and a review of implementation requirements to meet product cost, delivery, and manufacturing requirements.

An example will clarify the process. A company was working to reduce the cost of a low-cost television set, and cost analysis showed that the cabinet was very high-cost relative to other function areas of the product so they decided to apply the Blast, Create, and Refine technique to the problem.

In the Blast phase, after considerable discussion it was decided that the basic function was *contain components*. The first question in the Blast phase was, "What is the lowest-cost way to *contain components*"? A paper bag, cardboard box, paper cup, and others were among the answers. The paper bag was the lowest-cost alternative.

The next question was, "Will a paper bag satisfy the requirement?" "No, it is not rigid, will not satisfy fire requirements, etc." Now, each element must be taken separately, and the question, "What do we have to do to make it work?" was asked. To make the bag rigid it could be treated with a chemical. It could also be made fire resistant. What would that cost? Would it satisfy requirements? What would it look like? It looks like a waste basket. Could we use a waste basket? We might. Let us see if we can get a waste basket to serve our purpose.

Waste basket manufacturers were contacted and agreed they could supply a modified waste basket to satisfy all requirements at a fraction of the cost of a television cabinet. The result was better than a 50% cost reduction.

This method is admirably suited to brief discussions of product problems. Define the function. Determine the lowest cost way to provide the function and build on the basic function.

The use of this technique can be very painful to the originator of the original design. He or she has expended considerable effort to arrive at a solution and may look upon the resultant recommendation as criticism. This is an important reason for having the responsible person participate in any effort to change a product or process.

Summary

There are many techniques to help to improve creative ability. In addition, there are courses, workshops, and seminars; however, the primary ingredient is your desire. You must motivate yourself to make the effort to improve your creativity. It seems to me that successful creativity requires three ingredients, which I call the 3As.

1. Awareness — Input — Available knowledge garnered over a lifetime
2. Attitude — Desire — Inner need to do something, motivation
3. Action — Output — Result

Creativity is a gift everyone has, but few use. Why is it not used? Some people fail to realize they can be creative. Others do not see the need for it and do not feel it is worth the effort to work on improving their creative ability. Unfortunately, these people do not realize the meaning and significance of creativity. Most of our personal accomplishments in life depend on our creative skills. It is an ability we cannot do without if we wish to improve those things that surround us and make up the world we live in. Creativity can be ours; we need only to reach out and grab it.

References

1. Laderman, J., Brain Tune-up Suggested, *Detroit News,* Mar. 1, 1979.
2. Restak, R., *The Brain,* Bantam Books, New York, NY, 1984, 266.
3. Haefele, J. W., *Creativity and Innovation,* Reinhold Manufacturing Reference Series, New York, NY, 1962.
4. Heppenheimer, T. A., What made Bell Labs great, *Invention and Technology,* 12, 1, 46, 1996.
5. Adams, J. L., *Conceptual Blockbusting, A Guide To Better Ideas,* 3rd ed., Addison Wesley Publishing Company, Reading, MA, 1986.
6. Fry, W. F., Mind over mattress, *Flight Time Magazine,* Sept. 1974.
7. Good, P., *The Individual,* Time-Life Books, New York, NY, 1974, 143.
8. Hall, P., Dr. Noise, I. D. *Magazine,* May 1997.
9. Nietzsche, F., Quotable Quotes, *Readers Digest,* Nov. 1997.
10. Garfield, P., *Creative Dreaming,* Ballantine Books, New York, NY, 1985.
11. Osborn, A. F., *Your Creative* Power, Charles Scribner's Sons, New York, NY, 1949.

THE SUM
OF THE PARTS —
A PRACTICAL METHOD

IV

Chapter 12

Value Engineering: A Total System

Introduction

In Chapter 2 a number of management systems that have been popular in the past as well as some of the latest systems were briefly discussed. The intention was twofold. The first was to highlight some of the methods that have been popular over past years as a basis for illustration, and the second was to show that none of the systems is applicable in all cases, as shown in the Management Systems Toolbox. However, each system, whether still in favor or not, has elements that are crucial to problem solving, organization or personal development, and project design and development.

In Section 2, The Economics of Profit, the four elements of profit were discussed and their relationship explained. Cost is crucial to any business and to a wide variety of decision-making techniques. Understanding that the content of what people call cost can vary over a wide range, and the necessity to know the content to establish a basis for decision is important. The difference between cost, value, and the relationship to quality is also important, and when function is included in the package the range of opportunities for improved products and services at lower cost expands dramatically.

When working on any project, it is necessary to relate to a number of people of different backgrounds. In Section 3, The Human Element, several human factors were discussed. In today's society, working in teams is becoming the norm. Collecting information and understanding people's answers frequently requires some knowledge of what may be motivating them and what their value system may be. In addition, it is important to understand that creativity is more difficult for some people than others, and it frequently requires the application of a range of techniques to help them to break down the constraints that they may have applied to their creative ability over the years.

These technical and human factors are important to success in today's society whether in engineering or management. Although people are expected to perform successfully in groups, very few have had any training or background in these critical areas. The intention here is not to make anyone an expert in these fields but to point out that these technical and human factors exist and have a strong impact on the success of any enterprise and the people in them. It is hoped that we will have created enough interest in them to cause you to look more deeply into each subject to obtain greater benefit from your knowledge and skills.

All of these factors are brought together in the Value Engineering process, and an understanding of them can contribute to the success of a project. The knowledge and skill of the team leader in guiding and directing the team through the process not only expedites the process but increases the benefit of the result.

Function Analysis, A New Idea

The Value Analysis management system has been with us for a long time. The concept was developed by Lawrence D. Miles[1] while he was employed at the General Electric Company during the 1940s. The actual birth of Mr. Miles' concepts as a total system was in 1947 when the term *Value Analysis* was first used. Its acceptance at the time was so enthusiastic and success was so dramatic that the U.S. Navy Bureau of Ships became interested in 1954 and the term *Value Engineering* (VE)[2] was coined, since their primary function was engineering.

We do not see any difference between the application of Value Analysis methods to either analyzing an existing product or service or applying the process during the development stage. Since engineering can be defined as the science by which the properties of matter and energy are made useful to man, we have elected to use the term *Value Engineering* (VE) as synonymous with *Value Analysis* and the abbreviation VE. In addition, this term is recognized worldwide.

A professional society, the Society of American Value Engineers, (SAVE) was organized in 1959 to promote the knowledge and application of Value

Analysis/Engineering (VE). Since then, VE has been accepted worldwide as an extremely effective system to improve products and services for cost and productivity improvement. Recently, the society changed its name to SAVE International to better reflect this worldwide influence.

Unfortunately, the terms Value Analysis and Value Engineering seem to mean widely different things to different people. Many think that VE is simply the same old cost reduction system we have used for years with a new name. This is not the case. VE is an entirely new philosophy that is based on two premises.

1. Conventional approaches to problem solving stifle the imagination by restricting one's thinking to existing objects and methods.
2. Concentrating on function requirements, or what the product does rather than what it is, offers the maximum opportunity for creative solutions to problems.

There are many definitions for the system; however, Value Engineering is the only management system that is based on function analysis, or what the product does rather than what it is. The intention is to provide an entirely new viewpoint to allow new ideas that previously had not even been thought of much less considered. The method tends to cause a chaotic situation at the start because the project with which everyone was familiar changes to an entirely different entity. The substance of the project appears to change from a concrete being to a nebulous series of functions. However, the methods to reassemble the confusion are well defined and offer the opportunity for substantial improvement well beyond previously considered standards. The objective is to force people to think differently.

Definition

The SAVE International society defines Value Analysis/Engineering as follows:

Value Analysis/Engineering is the systematic application of recognized techniques that

1. Identify the functions of a product or service
2. Establish a monetary value for the functions
3. Provide the required functions at the lowest overall cost

These steps are implemented by following a well-defined Job Plan and at the same time being aware of the major elements covered in all of the preceding chapters.

To demonstrate the process we will carry an example of an actual product that has been successfully implemented through all steps of the Job Plan to implementation of the recommendations in the total assembly. In addition, we will follow the project to the next product redesign to illustrate how the long-range recommendations can be applied for future benefit. The overall program from the time of the original design to the future product implementation until the product was eliminated from production covers a period of about 25 years.

The two basic principles in VE are function definition and analysis and the Job Plan. Function definition was covered in detail in Chapter 2. Everything flows from these two principles. Of all the management programs past and present, these two principles are exclusive to VE. In fact, if you do not perform function analysis and definition you are not applying VE.

The Job Plan

The job plan is the second most important element of VE. It is a formula for action that leads a team through the complete VE process from start to finish. The Job Plan is an extremely important part of the process. It creates a structure that leads a team through the process and at the same time, compensates for the peculiarities of the individual. This compensation forces the application of analysis and creativity at the proper place in the process, thereby, tending to guide the team in the application of analysis (left brain) and creativity (right brain) functions.

To achieve maximum benefit from the Job Plan it must be rigidly followed. This may seem a paradox, however, rigid adherence to the plan offers ultimate flexibility in thinking. A typical Job Plan and one that we follow is in Figure 12.1. It is divided into six phases. However, each step is divided into subsets as shown. You may be familiar with a different plan than that shown, however, upon close examination you will probably find that only the nomenclature is different.

One of my associates is a retired admiral. At one time in his career, he was training a group of young midshipmen in ship handling. After some time he became so well known to the engineers that they knew who was handling the ship just from the engine room signals. As the ship was approaching the dock he would signal so many degrees on the helm and so many turns on the propeller. He never changed his pattern. As a result, he became well known throughout the fleet from his docking

THE VE JOB PLAN

1 Information Phase
     o Orientation
     o Determine Project Cost
     o Set Goal For Achievement
     o Define Functions
     o Construct ARGUS chart (FAST Diagram)
     o Pinpoint the problem, Define Targets For Opportunity
2 Creative Phase
     o Develop Alternatives
3 Evaluation Phase
     o Screen and Evaluate All Ideas
     o Identify Concepts
4 Planning Phase
     o Develop Concepts
     o Plan Recommendations
5 Reporting Phase
     o Organize Recommendations
     o Recommend Action
6 Implementation Phase

Figure 12.1   The Job Plan

pattern. Once a young midshipman asked him why he never changed. The reply was that he maintained his pattern so that he didn't have to think about where he was in the procedure; he knew exactly where he was at all times. This made it possible for him to react instantly to any emergencies.

By following the Job Plan you will know exactly where you are in the procedure. You will not have to worry about whether or not you have completed some phase of the project. You will know. This will free your mind for the specific task at hand, whether it is creative or analytical or developing a plan to sell your recommendations.

Information Phase

The Information Phase is critical to defining the problem. This is the longest and most complex step in the VE process and takes about half of the time spent on an entire project study. All of our analytical powers must be used to study, analyze, and define the data package. In addition, the functions of the project must be defined and evaluated to pinpoint the Targets for Opportunity (TFO) or the areas in the project to obtain maximum benefit from creative effort.

The Information Phase contains elements of the process exclusive to VE and critical to a successful project. Included here are goal setting, function definition and analysis methods, the Argus System of the Function Analysis System Technique (FAST), and function costing and evaluation. Completion of the Information Phase clearly and precisely defines the problem and indicates the potential for achieving the project goal.

The Information Phase defines the functions that form the basis for creative questions and provides an indication of the potential benefit for each function.

Before a VE project workshop is begun to analyze a project, a team must be selected, a workshop time set, and all participants must commit to full-time participation. In addition, a responsible person must be assigned the task of organizing the data package outlined in Figure 10.5, (Chapter 10). This is usually the responsible design engineer, manufacturing person responsible for the project or in the case of administrative or community affairs projects the person responsible for implementing the project recommendations.

The first step in the Information Phase is the orientation. This is where the team thoroughly analyzes the data package and becomes aware of the information available to them. This is a good time to get any additional information that the team feels may be required.

The second and third steps are to determine the project cost and set a Goal for Achievement (GFA). Project cost is thoroughly discussed in Chapter 2, Cost and Its Elements. However, it is well to point out that in the case of administrative or community projects cost is not always clearly defined. In many, if not most cases, cost in these types of projects may be in the form of time expended and may be in hours, days or weeks. It is also possible that the cost will be a combination of time and money from several sources such as federal and state grants and contributions. This information must be determined before the workshop team meets to begin to apply the VE process to save time and assure a smoothly conducted program. Although every effort may be made to obtain all of the necessary information, missing or required information may become apparent during the orientation. Once the project cost is determined the GFA must be set as a target to work toward.

In today's world, this GFA should not be less than a 30% reduction in the project cost. In some cases market and competitive conditions may indicate the goal should be 50% or higher, if you intend to be in business for the next 5 years or longer. This is especially true since history indicates that only about half of the proposals recommended are implemented on a timely basis. It is also important to recognize that studies have shown that the higher the goal the greater the performance will be, even though the group may be quite unhappy working toward the goal.

The fourth step is to define the functions, and the fifth step is to create an Argus Chart. The Argus Chart is a graphical representation of the entire project. Function definition and the Argus system go together. Construction of an Argus Chart is an essential aid in ensuring that the functions have been properly defined and that nothing has been overlooked. The next step is to cost the functions in the diagram and evaluate them in some measurable medium such as dollars, time, or hierarchy such as need or importance. Next, each function must be evaluated based on function cost vs. function value.

The sum of all of the function values will indicate the total system value. At this stage, the team participants should be optimistic in their evaluation of the potential for improvement. An optimistic evaluation will provide a better comparison for analysis and guide for future effort. The evaluation will be the lowest cost to provide the project functions based on the judgment of the team. This total value should be compared to the Goal For Achievement. The comparison will clearly indicate the potential for achieving the goal by identifying the functions that offer the greatest potential improvement. There will be times when this comparison will indicate a greater potential for achievement than originally estimated. There will also be times when the comparison will indicate that, in order to achieve the goal, a complete system redesign will be required. However, if this is the case, the cost-value differences will clearly indicate the function areas that must be improved for maximum benefit.

Functions that show the largest difference between the function cost and function value are the Targets for Opportunity (TFO) and form the basis for creative questions for the second phase of the Job Plan, The Creative Phase. The Information Phase defines the problem and its scope.

The TFO are those areas of the project that we expect to yield maximum benefit from our creative effort. If we have done a good job of defining the functions, they form the basis for the creative questions and in some cases, may be the creative questions themselves. This tends to assure us that we are solving the real problem and not just a symptom of the problem. Only after we have the creative questions are we prepared to enter the second phase of the Job Plan, the Creative Phase, where we will develop as many ways to satisfy the creative questions as possible.

Creative Phase

The Creative Phase should be approached with a completely open mind with the sky as the limit. It is not always easy to turn on the creative process. Not only have most of us tended to stifle our imagination to fit into society, but along the way we have created a number of effective mental blocks. These blocks must be broken down for effective creativity. The first step in doing this is to completely separate the Creative Phase from the analytical process and then make use of one of a number of techniques designed to foster creativity and the creation of ideas. A technique we have found most effective and easy to use is Brainstorming. It is always a good approach to start the Brainstorming session with a few creative games to loosen everybody's thinking apparatus and then to explain the rules and proceed until an adequate number of ideas have been generated.

The way to start a brainstorming session is to list the functions identified in the Information Phase as targets for opportunity. These functions are those that show the largest difference between function cost and function value. Out of 30 to 40 functions it may be that only 5 or 6 offer the opportunity for project improvement and are worthy of creative effort. Each function should be "brainstormed" until no new ideas can be listed. No evaluation or discussion is allowed at this stage; that will come later. All ideas must be written down to assure that nothing has been overlooked. A good way to do this is to list the ideas on large tear sheets that can be posted on the wall for everyone to see during the Evaluation Phase. Remember we are looking for quantity, and every idea, no matter how ridiculous it may sound, may trigger a winning idea in someone else's mind. Several hundred ideas are often listed.

Evaluation Phase

In the Evaluation Phase we return to analysis and discuss and question each and every item on the Brainstorm list. The objective here is to find something that will form the core of a concept that can be developed to provide a solution to the task. If we can't find it here we will probably not successfully achieve our Goal for Achievement.

The first step in the evaluation process is to read through the list and eliminate any items that have absolutely no potential use. At this stage do not get bogged down in discussion. However, the intent is not to eliminate items but to keep any that may have the slightest potential for benefit. After this first quick screening, go through the list again looking for all and any that may provide the core of a concept for further development. Always look for more than one concept to provide insurance for unforeseen circumstances.

It will be found that the result of the first screening will classify the random ideas into several categories such as materials, fastening methods, structures, etc., or in the case of processes, storage methods, urgency of methods, etc.

Planning Phase

The Planning Phase grows almost automatically out of the Evaluation Phase; however, it is now necessary to begin to organize the various ideas and concepts into practical project solutions. In the course of the Planning Phase it is necessary to consider the effect of the recommendations on the product and the organization. It will be necessary to forecast the project cost, to identify who will be affected by the changes and what they will be expected to do to successfully implement the project. It is also necessary to determine the cost of any new equipment, organizational effect and any other possibilities. Adequate and even superior preparation in this phase of the Job Plan is essential to success in the overall project. The worksheets have been designed to aid in preparing a complete and effective recommendation and to ensure that nothing has been overlooked. The example later in this chapter, will demonstrate how the worksheets are used in the course of a project, and a complete set of blank worksheets is include in the Appendix.

Reporting Phase

Finally, the time has come to present your recommendations for approval. Although this is the last step in the workshop it is not any the less important. If ideas and recommendations are not presented effectively, an excellent recommendation can be lost. It should be remembered that an excellent recommendation can be rejected by a poor presentation, but it is very difficult to sell a poor recommendation, no matter how good the presentation may be.

The recommendation is a sales presentation, and any good salesperson knows that he or she must ask for the order. Remember that the last thought you must leave with the audience is the action you want them to take. One of the greatest disasters that can befall a team's presentation is to do an excellent job and present an excellent set of recommendations in a first-class presentation and then not specify the action required from the decision makers. However, we have found that in at least 90% of the cases teams do not design their presentations to ask for the order. For example, if approval is required to begin design work, ask for the approval to get the funds to start work on a specific date. No matter how many times people are reminded of the importance of this request they seem reluctant to do it. Don't lose an effective proposal because your audience doesn't know what they have to do to obtain the prize.

Implementation Phase

Although the Implementation Phase is not a part of the project workshop, every project must have a champion. This is one of the main reasons for having the person responsible for the project as a member of the team. Of course, the main reason is that when he or she takes part in the workshop the person becomes part of the solution and should be interested in seeing the recommendations implemented. He or she is also the most likely candidate to be assigned as the project champion to follow development through to successful implementation. In many cases, depending on the stage of the project development, he or she can see that recommendations are incorporated directly into the project program.

This is a very brief outline of the activities that must be conducted in each phase of the Job Plan. Each phase deserves a complete section of its own. In reality, however, there is only a philosophy that can be followed because each project is different, especially in that the people involved are different and carry a different set of values from any other project team.

Job Plan Examples

There are a number of different job plans in the literature. The one I have illustrated here is six steps, simple, easy to remember and understand. However, as explained, the sixth phase is beyond the workshop scope. No matter how many steps there are, the process is always the same, analysis, creativity, evaluation and development. Several different job plans are shown below for illustration. In many cases, the phases may be the same but have different names.

MILES
1. Orientation Phase
2. Information Phase
3. Creative Phase
4. Evaluation Phase
5. Planning Phase
6. Recommendation Phase
7. Implementation Phase

DEPARTMENT OF DEFENSE[3]
1. Orientation phase
2. Information Phase
3. Speculation Phase
4. Analysis Phase
5. Development Phase
6. Presentation and Follow-up Phase

FALLON[4]
1. Preparation Phase
2. Information Phase
3. Analytic Phase
4. Creative phase
5. Evaluation Phase
6. Presentation Phase
7. Implementation Phase

MUDGE[5]
1. General Phase
2. Information Phase
3. Function Phase
4. Creation Phase
5. Evaluation Phase
6. Investigation Phase
7. Recommendation Phase

VALUE ANALYSIS INC.[6]
1. Information Phase
2. Speculation Phase
3. Analytical Phase
4. Planning Phase
5. Execution Phase
6. Recommendation Phase

GERMANY, VDI Spec. VA/E 69-910
1. Preparatory Measures
2. Determine Existing Condition: Defining functions and function costs
3. Verifying Existing Conditions
4. Generating Solutions
5. Verifying Solutions
6. Proposal and Implementation

FRANCE, Specification NF × 50 152 (August 1990)
1. Orientation
2. Data Collection
3. Function and Cost Analysis
4. Gathering and Reviewing Solutions
5. Building on Ideas
6. Present Recommendations
7. Implement Recommendations

JAPAN[7]
1. Information Collection Stage
2. Function Evaluation Stage
3. Creation Stage
4. Schematic Evaluation Stage
5. Detailed evaluation Stage
6. Proposal Preparation Stage
7. Proposal Follow-up Stage

The Value Engineering Process

An example of the application of the VE system to a simple product will illustrate the process.

In the early days of automotive air conditioning, the system was a very expensive option and was installed only on a few luxury models. As development improved the operation of the equipment and costs were decreased, the accessory was more and more in demand. However, in the early stage the limited volume dictated low-cost tools as a practical investment. As a result, the distribution outlets, known as spot coolers, were designed and manufactured as die-cast assemblies. Although the tools were relatively inexpensive, the parts were rather high cost. Each spot cooler cost $3.80, and there were four units required for each vehicle.

Once the public experienced the comfort of air conditioning, the demand increased rapidly and made lower costs essential and practical. The first step was to change the material to plastic. Although the tool cost increased substantially, the part cost dropped by 73% to about $1.00. Since there were still four units per car, this was a sizable cost reduction of $11.20. Although the cost was reduced substantially, from a function standpoint the product was exactly the same as the die-cast unit with very few exceptions.

To further improve the overall cost of the system and at the same time see what VE could do, a task force was assigned to take the new plastic unit as a workshop project. The team was made up of the manager of the department that had designed the original unit, a materials specialist, a styling engineer, and two cost specialists to educate them in the process. The team was composed of a group that had all of the necessary information to solve the problem.

Orientation and Goal for Achievement — Step 1

The first step was to provide a one-hour orientation to explain how the workshop would be conducted and to give a brief explanation of the process. The team was then asked to review and become thoroughly familiar with the information package, determine the specific product cost, and set a goal for achievement.

At this stage in the project the team should set a "stretch" goal. From past experience and studies made to evaluate the effect of goals on performance, it is known that the higher the goal the better the performance is likely to be. However, the team may not be very happy working toward the goal. It has been shown in studies that when a small group was asked to come up with 10 ideas for a new product they came up with 15. However, when the group was asked to come up with 20 new ideas they came up with 18. Although they did not make the goal their performance was better. When the same group was asked to come up with 30 new ideas they developed 25. Again, they did not make the goal but their performance improved. Repeated experiments have shown that the higher the goal the higher the performance will be. However, the higher the goal the more stressful the effort will be.

In setting the goal for achievement the team can base its target on present cost and on market position. One project we had was based on vacuum tube technology; however, a competitive company had a solid state unit that could perform the same functions at half the cost. The team considered setting its goal at 50% reduction but then realized that the competition might be in a position to raise the ante, so they set a goal of 65% reduction and made it. There are many cases where competition may set the goal for you. When this is the case, remember that if you match the competition cost they may be able to incorporate new changes and reduce their cost further.

There are also cases where management will set the goal for you. In addition to reducing the cost they may also add features that will force the need for cost reduction. Then there are the facts of life where your customer is requiring a 4 to 5% cost reduction every year of your contract. If the contract runs for five years this means you are agreeing to reduce the cost by about 25% within five years, so why not set the goal and work toward achieving the goal as soon as possible.

In the case of the spot cooler the design engineer said that there was no way the cost of the present unit could be reduced further; that $1.00 was a good buy. This was true, however, we said, "Let's look at the functions and see what we get." We also pointed out that our experience indicates that at least one third of the cost in any product is unnecessary and can be eliminated without affecting the required functions of the product. Based on this and the desire to achieve a successful result, the team set the GFA at 50%.

This GFA is your target. It should be as optimistic as possible. Remember that it may only be possible to incorporate half of the recommendations into the current product and that it may be necessary to develop a new product to obtain all of the potential benefit from your recommendations.

The team now completed Worksheet No. 1 as shown in Figure 12.2.

Define the Functions — The next step is exclusive to the VE process. The team must define the functions of the product. This is quite difficult because it requires an in-depth analysis of what the product does rather than what it is. Remember, the customer does not buy a product; he buys what the product does for him, the function. Although this requires a different way of looking at the product and is critical to a successful workshop result it is not important to strive for properly defined functions at this stage of the project. The first step is to begin to develop the two-word technique, which is difficult enough. Proper functions will develop as the project progresses and the team becomes more familiar with the system.

Start the function definition process by looking at the functions of the entire assembly. After it is believed that all of the functions have been defined, define the functions of each piece in the assembly, even the nuts and bolts. At this stage there will be duplication of functions already defined. Don't let this bother you. It is easy to scratch out the duplicates, and we want to ensure that we haven't forgotten anything. However, the function should be defined only once. For example, in the spot cooler there are several places where transmit torque or transfer force are defined but the function is listed only once. The intention is to consolidate all costs for each of these and other functions under the one definition. At this stage there is frequently the fear that this will cause some of the data to be lost. This will not be the case as will be seen in the next few steps of the process.

Define the functions on sheet 2. The worksheet for the example project is shown in Figure 12.3. It has been screened to eliminate duplication.

As noted above, several parts may perform the same function. We want to find out how much we are spending to perform each function. The fact that different components may perform the same function may lead us to the conclusion that they may not be needed.

Information Phase   Sheet 1

Project
Information   _X/XX/XX_ Date

Part Name		Drawing or part no
Spot cooler assembly	A/C and Heater	2936148
Used on (Name or Number)		Number required per assembly
All Carlines		one (1)
Team number	Workshop number and date	Task force and date
6		

Team Members	Department	Phone number
Jim Kress	A/C Elect & Mech.	6-4431
Stan Stahl	Cost Analysis	6 - 4806
Paul Drew	Cost Analysis	6 - 4398
Bill Cook	Styling Design	6 - 4977
Art Rose	Materials	6 - 4161

Present cost

Total cost	Cost Elements		
$1.00	Material $ 0.3427	Labor $ 0.2338	Burden $ 0.4235

Estimated annual production   250,000 Units

Operation and Performance

Assembly controls air by modulating flow from on to off and by directing air
in a horizontal or vertical direction. Modulation is performed by a door &
actuating lever; directional control is achieved by rotating vanes or vane
assembly. Assembly is activated by operator. Assembly consists of 18
separate pieces.

Worksheet No. 1

Figure 12.2   Information Phase

After the list has been screened, find the basic function. The basic
function is the one upon which all other functions depend. If it wasn't
necessary to perform the basic function it would not be necessary to
perform any of the others and the assembly may not be required. Read
down the list and put an x in the spot for the basic function or functions.
It is possible for a product or service to perform two or more basic
functions; however, each function usually applies to a different set of laws

1. *Information* **Phase**   **Sheet 2a**

Determine Scope
Identify Functions   ⟋XX/XX␣Date

Project Name	Drawing/Part No.
Spotcooler Assembly	*2936148*

What is It?	**Scope includes**
--	

Other limits of project.	**Scope does not include**
None	

What does It do?
Modulates and controls and directs air

List all functions without constraints

	Verb	Noun	Basic	Second	Remarks
1.	*controls*	*air*	x		
2.	*contain*	*assembly*		x	
3.	*attach*	*assembly*		x	
4.	*improve*	*appearance*		x	
5.	*modulate*	*air*	x		
6.	*vary*	*opening*		x	
7.	*apply*	*torque*		x	
8.	*direct*	*force*		x	
9.	*limit*	*rotation*		x	
10.	*direct*	*air*	x		
11.	*rotate*	*louver*		x	
12.	*support*	*louver*		x	
13.	*prevent*	*distortion*		x	
14.	*rotate*	*vane*		x	
15.	*increase*	*convenience*		x	
16.	*interlock*	*vanes*		x	
17.	*control*	*friction*		x	
18.	*retain*	*vanes*		x	
19.					
20.					
21.					
22.					
23.					
24.					
25.					
26.					
27.					

Figure 12.3   Worksheet No. 2

(for example, electrical or mechanical; physical or chemical). In the case of systems they may be design and manufacturing, or educational, or operational.

A way to check whether the selected functions are basic or secondary is to apply the following set of questions. On the function list, three functions were selected as apparent basic functions — control air, modulate air, and direct air. Now ask this question; only yes or no answers are acceptable. "If I didn't have to *control air* would I still have to *modulate air?"* The answer is no. Now ask, "If I didn't have to *modulate air* would I still have to *control air?"* The answer is yes.

What this has told us is that *modulate air* is dependent upon the need to *control air.* If there is no need to *control air* there is no need to *modulate air.* The same questions applied to *direct air* will produce the same result. There would be no need to *direct air* if it wasn't necessary to *control air.* Therefore, *control air* is the basic function. Both *modulate air* and *direct air* are dependent upon the need to *control air.* All of the remaining functions are called secondary functions and are subject to modification or elimination depending on their cost and value. Secondary functions theoretically make the basic function work or sell better.

Function Analysis and Evaluation — Over the years a number of techniques have been developed to aid in analyzing and evaluating functions, to help to clarify problems and to identify unnecessary cost. There will be cases when the very simple techniques may be all that is required. In many cases, just asking why something is being done and requiring a two-word answer in function language will cause a person to see a problem in an entirely new light. In fact, they may realize that their problem is not what they thought it was.

Six basic methods are listed here. The order in which they are given has no particular significance. Practice will develop knowledge and skill in their use; however, in most cases we use Technique 6. The Argus system. Although it is the most complex it is also the most straightforward and helpful and usually forces the development of information that was not previously considered.

The six techniques are as follows:

Technique 1. Identify and Evaluate the Function
Technique 2. Evaluate Principle of Operation
Technique 3. Evaluate Basic Function
Technique 4. Theoretical Evaluation of Function
Technique 5. Input Output Method
Technique 6. Argus System, Function Analysis System Technique (FAST)

Technique 1.   Identify and Evaluate the Function

This is a simple technique that asks the question, "What must the part or assembly do?" It applies to all projects and requires a clear determination of all use and esteem functions. Each function must be expressed in the two-word verb–noun function language. We firmly believe that if you cannot define the functions in the function language you need to better understand the project. Worksheet No. 2 has been designed to aid in developing this technique.

Technique 2.   Evaluate Principle of Operation

This technique is essentially the same as Technique 3 except the emphasis is on principle of operation. The technique requires a detailed examination of physical laws and effects upon which the function could be based to allow a simpler more reliable and less costly operation. For example, to provide data on auto engine temperature, a *transmit information* function based on laws and effects that apply to heat might be replaced by one based on magnetic principles.

This approach has broad application on new items and can be a useful tool for advanced design and research operations. For example, in the design stage the decision to provide mechanical, electrical, pneumatic, or other means to provide automatic temperature control would have a major effect on system design.

Technique 3.   Evaluate the Basic Function

This technique imposes the strictest discipline and requires the acceptance of a forcing assumption — "only basic function has value." The assumption is made as a mental step to force thinking to search for new and simpler designs or ways to do things, that will provide the basic function in a way that the least number of secondary functions will be required to make it work and sell.

This technique is best applied to assemblies; however, it can be modified for use with single parts. The Blast–Create–Refine technique, as described in detail in Chapter 11, Creativity, is an example of a special case of this technique. The value, as is developed here, is the combined result of individual judgment, creativity, and past experience as to what a function should cost, based on the work it performs and the way it could be done.

There are many variations of this technique. One is to expand the scope of the study and eliminate imposed functions by revising each listed function determined in Technique 2, and asking the question, "Is this function performed this way as the result of the basic design concept?" Redesign to eliminate imposed functions means expanding the scope, thereby causing adjoining components to dictate new limits. Some large savings can be made as a result of the skillful use of this technique.

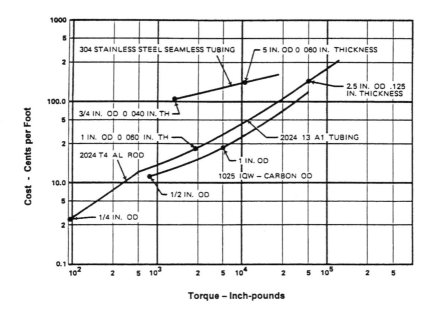

Figure 12.4   Value Standards for Transmit Torque

Technique 4.   Theoretical Evaluation of Function

Techniques 3 through 6, based on skill, inventiveness, and experience as the means for determining the value of a function, have proven to be widely accepted approaches. The theoretical evaluation of function places a precise value on a function by using appropriate mathematical relationships. It applies to measurable parameter functions only such as create heat, resist bending, as opposed to functions that improve appearance, maintain decor, enhance status, etc. In *Value Standards*,[8] Roy Fountain developed a mathematical method to scientifically determine a function value.

In the graph in Figure 12.4 below, the cost in cents per foot has been charted against the torque carrying capacity for various materials. This graph instantly highlights the cost required to satisfy the function "transmit torque."

This approach takes VE from an art to a science and opens the door for value research. While the basic concept is still the same, equating cost to function, a considerable grasp of basic value techniques and mathematics is required. In addition, it is necessary to continually update the databank.

Item	Input	Function	Output
Transformer	6 volts D. C.	Amplify Voltage	12 volts D. C.
Hot water heater	cold water	increase temperature	Hot water
	Power	Convert energy	Heat
Pipeline	Fluid	Transmit Fluid	Fluid
	Energy	Convert energy	Pressure

Figure 12.5   Input–Output Technique

Technique 5.   Input–Output Method[9]

This technique is useful in highlighting the basic function of a product by viewing it as a black box item which receives certain inputs and transforms them into outputs. These inputs and outputs are not functions and therefore do not have to be defined in two-word terms. The function itself is a result of the input and causes the output; hence, the function is positioned between the input and output.

In Figure 12.5, 6 volts DC is the input to the transformer and 12 volts DC is the output. The function that fits between the input and output is "amplify voltage." Additional examples of this technique are listed below.

It should be noted that any item can have more than one input and output and unless inputs are transformed into outputs, the item has no value. Since function is the key link between input and output, this is equivalent to "only function can have value."

Technique 6.   Function Analysis System Technique — The Argus System

There is an old adage that says, "A picture is worth a thousand words." A number of writings could cause us to question this wisdom, such as the Lord's Prayer, Lincoln's Gettysburg Address, and the Declaration of Independence. However, it has been proven time and again that visual aids tend to substantially increase understanding.

Julius Caesar drew diagrams in the sand with a stick to aid his generals in battle. In the 17th century shipwrights used detailed models to illustrate their designs. Today, we use maps to aid in travel or drawings to guide mechanics or builders in the construction of a product.

Visual aids bring the sense of sight into the process of communicating an idea. Can we bring this visual sense into the analysis of a problem? Yes, we can and do in the VE process by constructing an Argus diagram.

FAST is the acronym for Function Analysis System Technique. A FAST diagram is a logic chart that organizes the functions of a project and arranges them in a cause-and-effect relationship. Although we consider FAST diagrams to be a major addition to the VE process for any project, we have found them to be particularly useful in the analysis of organizations, operations, or a project that may not be clearly defined or about which there may be widely differing opinions to the point of controversy. Construction of a FAST diagram tends to pull together the thinking process of a group to create a dynamic, enthusiastic team. The concept is simple. However, creating the diagram can be difficult and frustrating because it forces thinking about the project.

In the earliest days of VE the team would define the functions of the project then use a number of techniques to organize them for evaluation. These techniques ranged from random selection to one or a combination of the techniques described above to precisely define the specific areas to apply creative effort to achieve maximum benefit. At the 1965 SAVE International Conference at Boston, Charles Bytheway[10] presented a paper, "Basic Function Determination Technique." In this paper Bytheway pointed out that every function satisfies a cause-and-effect result. He then invented a way to illustrate this relationship that he called a FAST diagram or Function Analysis System Technique. Before this, I don't believe anyone ever tried to relate the functions in any way. I call this relationship the Theory of Function Relationships (TFR).

Although Bytheway invented the FAST diagram to illustrate this relationship it was found that several of the complexities in the original process were not necessary to obtain the benefit of the diagram. In addition, the method of construction of the Bytheway process was changed from control by the team leader to control by the team. Several benefits accrued from these changes, and the application of the diagram was expanded to consolidate the entire Value Engineering Job Plan into an organized unified system. Over time, several different methods for constructing the diagram have evolved and set our method apart from the others. We have named our method the Argus System, which results in an Argus Chart. Argus refers to the mythical being with many eyes who was able to see things from all sides. The method described here will be based on our understanding and development of the TFR over a period of years and thousands of successful project applications.

The Argus System

The Argus System makes use of the TFR. In addition, we strive to use the construction of the chart to bring the team together by having them construct the diagram based on their knowledge and understanding of the problem. The Argus System of FAST diagramming does this and in addition, provides a continuous link that ties the whole system together into a complete management system. However, we insist that the problem will tell you what it is. You cannot change the problem to satisfy your wishes by making it one you may think you know how to solve. Participants are forced to think deeply about the problem because the functions as originally defined may not fit the how and why logic of the system. The team must stretch its thinking, change its role, and do anything and everything possible to satisfy the diagram logic requirements. This forces insights that open the mind to a far better understanding of the project than anyone had before. The completed chart then becomes a graphic representation of the project and assures everyone that it thoroughly describes the task. It can also be used for a number of other factors in the future.

In many cases an Argus Chart is used to organize the functions of a product, but it can be equally useful in helping to understand abstract projects such as productivity analysis of technical, manufacturing, and administrative operations. Argus Charts have been used to design jobs, create new organizational structures, aid in career planning, and a myriad of other projects and usually result in more innovative methods to achieve the required objective.

Constructing of an Argus Chart — Construction of an Argus Chart is one of the things you learn best by doing. Construction depends on a few simple rules. However, application of the rules requires determination and discipline. The first step is to create a list of functions for the project. At this point it is not necessary to struggle for the best function definition. If you are applying the process for the first time, just developing the two-word technique is difficult enough. As you get more experience the two-word technique will become second nature. In addition, you will begin to get the feel for what constitutes a good function definition for your project. If you cannot define the function in two words, you probably do not understand the problem; review the project and think harder.

Define the Functions — When defining functions, start at the top. In other words start with the entire assembly or system. Next take each part or step and define the functions. There will be duplication which will be eliminated after the list has been completed. We want each function to be defined only once so that all costs for that function, no matter where they may be within the project, will be assigned to one function.

After the duplication has been eliminated, each function must be written on a one-by-two inch card or chip. While this is being done, other members of the team can prepare the worksheet. A typical worksheet could be 24 by 36 inches. The first step is to lay the sheet out as shown in Figure 12.6. It is extremely important that the how, why, and when designations are shown to ensure that the proper questions are asked.

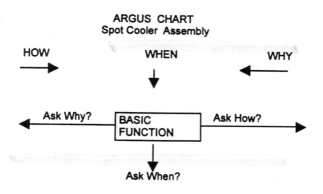

Figure 12.6   General Argus Chart Layout

Select the apparent basic function as discussed in Chapter 3, Function, and place this function chip near the center of the worksheet. This will make it possible to work in any direction depending upon the direction the chart construction leads.

Basic Logic Questions — Usually we know what we want done but we don't know the best way to do it. The first step in the use of the Argus System is to define the functions of the product, assembly, or service. The basic function will usually be among them but not always. If it is a relatively simple problem, this will not entail great difficulty as the basic function will show up during the chart development. If the project is very complex, the basic function will be the one, or usually no more than two, upon which all other functions depend. In other words, if the basic function is not needed there is no reason for any of the others.

There is a series of questions to keep in mind during the construction of an Argus Chart. They are shown in Figure 12.7. These questions should be used as they are required, but it may not be necessary to use all of them nor is there any particular order to use in applying them.

The two most useful questions are the primary path questions. HOW is this function actually accomplished, and WHY must this function be performed? These functions are used during construction of the primary path. The next most useful question is WHEN is this function performed? This function aids in pacing secondary functions on the chart and also helps to position a secondary family tree.

Define the Functions — Before starting construction we try to define all of the functions in the project. However, in most cases many of the functions as originally defined do not fit the logic questions or must be redefined. In addition, we may find that we did not define a necessary function so we must identify it and add a new function chip to the system.

BASIC LOGIC QUESTIONS

1.   **To find the higher order function**
     Ask the following logic questions:

     o   What am I really trying to do when
         I perform this function?
     o   What higher level function caused this
         function to come into being?
     o   Why must this function be performed?

12. **To find basic functions or to establish a family tree.**
     Ask the determination logic questions:
     o   If I didn't have to perform this function
         would I still have to perform the others?
     o   In this system, does the performance of this
         function cause this apparently dependent
         function to come into being?

3.   **To establish a primary path**
     Ask the primary path functions:
     o   How is this function actually accomplished?
     o   Why must this function be performed?

4.   **To locate secondary functions**
     Ask:
     o   When does this function happen?
     o   This function happens at the same time
         as what other functions?

Figure 12.7   Basic Logic Questions

This is a natural development in the construction of the system. What is happening is that more is being learned about the project. The chart construction is forcing in-depth thinking, and as a result new things are being discovered.

If the project is on an existing product or system, work with that system exactly as it is. Do not add things to the system that do not exist; if you do, confusion will result. If you are working with a new system or to invent a new product you will be able to define only a few of the functions and must use the logic questions to guide you through the system to add required new functions. An example of this type of system will be shown in the next chapter. Although the problem is different, the method to be followed is exactly the same.

Next, prepare one-by-two inch cards for each function. Pick out the apparent basic function and place it in the center of the prepared work-sheet and ask the question, "How is this function actually accomplished?"

Figure 12.8   Argus Construction 1

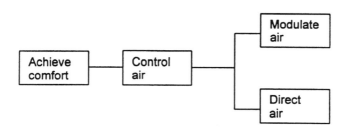

Figure 12.9   Argus Construction 2

The answer will give the lower-order function that has been caused by the basic function.

The answer should be among the function cards already prepared. If it is not, you may have found a new function, or one of the functions you have already identified should be redefined. Place this function to the right of the basic function. Now ask, "Why must this function be performed?" to determine the function to the left, or higher-order function. The answer should be the basic function. However, it may not be the basic function. It may be another of the functions already defined or it may again be necessary to define a new function.

Spot Cooler Analysis — Now let us apply the Argus System to the spot cooler problem. The basic function from the list in Figure 12.2 was determined to be *control air*. The logic question, "How is this function accomplished?" was asked to find the lower-order function, and "Why is this function performed?" to find the higher-order function (Figure 12.8).

How is air controlled? In this product, at this time the reply is *direct air* and *modulate air*. Both answer the How question. Do they both answer the Why question? Why is the air modulated? Why is the air directed? In both cases the answer is to *control air*. There must be two paths because the answer to the How question would not be satisfied with only *direct air* or *modulate air*. The logic questions are satisfied, and we can add the next diagram block (See Figure 12.9).

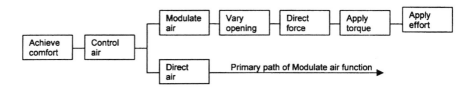

Figure 12.10   Argus Construction 3

In this system *control air* is performed by both *modulate air* and *direct air*. The present system cannot satisfy the function *control air* without both directing and modulating air. Therefore, the primary path must be split into two parallel paths.

Why do we *control air*? To *achieve comfort*. How do we *achieve comfort*? *Control air*. So the basic logic questions have been satisfied for the basic function *control air*. The basic function has been isolated, and the rest of the primary path functions can be determined. The function *achieve comfort* is a higher order (HO) function outside the scope of the project. However, the higher order function makes the basic function necessary.

Take one or the other primary path and continue to develop the chart. Do not jump back and forth from one primary path to another as this leads to confusion. These primary path functions become the framework for developing the complete Argus Chart. The How and Why logic questions have been applied to complete the modulate air primary path in Figure 12.10.

When these questions have been answered satisfactorily, the position of a known function is established within the diagram. In some cases a new function is discovered, and the primary path functions must be asked of the new function.

When the primary path for the *modulate air* function branch and *direct air* branches have been completed, there are a number of functions left over. A third consideration now becomes important. When do these functions happen? Normally, they happen at the same time as some primary path function. For example, when the *control air* function is taking place the assembly is contained and attached to a larger assembly, and style and design are being provided. Do not start to add When functions until the primary path has been completed, because confusion will result. Take one step at a time.

These functions line up vertically under or over the *control air* function and are connected with a solid line. They happen at the same time. Remaining functions are placed on the chart until all of the functions have been incorporated in the chart. It can then be considered to be complete as shown in Figure 12.11.

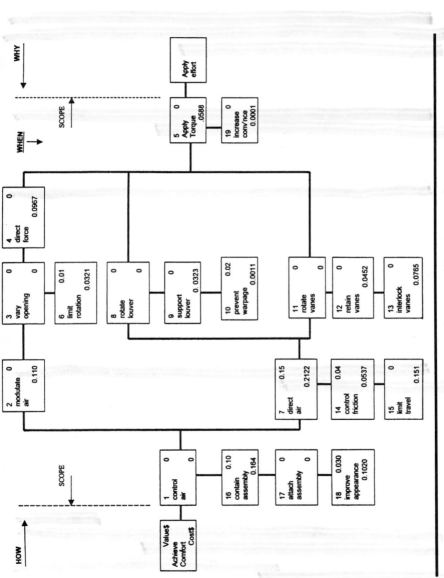

Figure 12.11 Function–Cost–Value. Spot Cooler Assembly — Argus Chart

Note that when functions are connected into the chart by a line to the top or bottom of the function box, primary path functions are always connected to the left or right side of the function box.

Be suspicious of too many *When* functions in a vertical line. In many cases a number of the so-called functions will be found to be actions, and they can be combined into a single function. It is also a good idea to get into the practice of checking every *When* function to see if a family tree develops. A family tree is a secondary path through the system. In many cases it may cover a different area of the problem than the basic functions.

Another test is to check the primary path to determine if all the functions are needed or if any can be eliminated. The way to make sure all of the functions are required is to test by removing functions individually to see if the system still answers the *How* and *Why* questions. If it does, the function may not be required.

When the diagram has been completed, the functions are numbered. Number the functions in the completed chart by starting with the basic function as number 1 and proceeding through the chart down the primary path. If there are two or more primary paths, follow the same procedure. This method will be an advantage in the function–cost–value analysis, which is the next step in the process.

Now that we know what the functions are, the cost of each function must be determined and added to the diagram. Total unit cost for the product assembly was determined to be $1.00. Costs are shown in the lower right corner of each function box in Figure 12.11. The value of each function will be added in the upper right corner of each function.

When this task is completed, the diagram becomes a very useful tool to show function relationships. The diagram is now a logical representation of the How–Why relationship between every function. It indicates clearly and concisely what is taking place in the project. The cost of the product can now be distributed to each function, and each function can be evaluated to determine if it is worth its cost. This is done by applying the tests for value. Methods for establishing function value are discussed in Chapter 4. The sum of the function values from this evaluation will be the value of the assembly. It will also be the lowest possible cost for the assembly and thus establishes the cost reduction target. The differences between the function cost and function values pinpoint the high cost areas where creativity can be applied for maximum benefit.

Target for Opportunity — The TFO will be the greatest difference between function cost and function value. In most cases there will be several functions showing a large difference between cost and value. Select these functions and arrange them by decreasing dollar amount difference. These functions will form the basis for the creative questions for creative action in the Creative Phase. In a large project with 30 to 50 functions, usually 6 to 10 functions form the basis for the TFO and will provide maximum benefit from creative effort. The first step in the process is function cost development.

Function Cost Development — In this spot cooler project the diagram shows that the *modulate air* function costs $0.11, or 11% of the total product cost. How was this cost developed?

To illustrate the function cost breakdown system, consider the thumb wheel. The cost chart, Figure 12.12 shows that the thumbwheel, item 12, costs $0.0957. The first step is to determine what the thumbwheel does. When the functions were defined, it was decided that the thumbwheel performed three functions, *improve appearance, apply torque,* and *limit rotation.* It is chrome plated only to improve its appearance, and a boss is molded onto the wheel to limit rotation. A check of the process sheets indicates that it costs $0.0501 to plate the item. So the cost to improve appearance is $0.0501. It was estimated that the boss required 10% of the material, and no other costs were involved. So 10% of the remaining cost or $0.0044 was identified as *limit rotation.* The balance, $0.0412 was then the cost of *apply torque.* The total of the function costs is the same as the item cost $0.0957. Notice that the cost of the basic function of the part, apply torque, costs less than its secondary functions.

The same techniques are applied to every part in the assembly. In some cases educated opinion must be used; in other cases data may be available to help in the decision. However, although every effort must be made to do as creditable a job as possible, accuracy doesn't have to be in four decimal places. There will be a great deal of opinion expressed, but we are only looking for guidance at this stage. In many instances someone may remark, "That's all it does and that's what it costs!" This is a clear indication that cost visibility is improving.

The easiest way to perform this cost distribution task is to select a part and read across the chart identifying every function performed by the part and put a small check in each box. Now, examine the selections and determine what functions may be performed for no cost and eliminate them by placing a dash in the appropriate box. Next, determine if there is factual data regarding any of the remaining functions such as painting or plating cost. Once this has been done, the remaining cost can be distributed by percentage.

The complete cost breakdown for the spot cooler is in the Function–Cost–Value worksheet, Figure 12.12. The cost should now be added to the Argus Chart in the lower right corner of each function. The chart now tells what the product does, why it does it, when things happen on

Cost Function Analysis

Cost Function Work Sheet — Team No. 6, Rev. N. 1969, Date Rel. Assembly – Part Name: SPOTCOOLER ASSEMBLY; Assembly – Part No.: 2936-148

Manufacturing Cost $ 1.00

Item No	No Req	Name/Action	Raw Material $ 0.3427	Labor $	Burden $	Other $	Unit Cost $ .2338	Total Cost $ .4235	CONTROL AIR (V I N)	MODULATE AIR (2)	VAC; METERING (3)	DIRECT FORCE (4)	APPLY TORQUE (5)
1		2936148 MOVABLE ASSY	.0155	.0682	.0862	.0311	.1910	.1910	.0331			.0550	.0176
2	1	2936174 UPPER ASSY	-	.0626	.0795	.0298	.1649	.1649					
3	1	2936480 RETAINER	.0028	.0029	.0105	.0172	.0213	.0213					
4	2	2622346 SPRING PIN	.0039	-	-	.0004	.0043	.0086					
5	1	2936483 VALVE CENTER	.0013	.0027	.0072	-	.0142	.0142					
6	4	2936174 VALVE	.0012	.0027	.0072	.0031	.0143	.0572					
7	1	2936186 TIE BAR	.0007	.0021	.0029	.0018	.0075	.0075					
8	1	2936188 SPRING	.0007	.0016	.0022	.0013	.0058	.0058					
9	1	6027529 VALVE RETAINER	.0022	-	-	.0001	.0023	.0023					
10	1	4027530 DRIVE KEEN PROT	.0022	-	-	.0001	.0025	.0025					
11	1	2936496 RETAINER	.0009	.0011	.0029	.0003	.0052	.0052					
12	1	2936170 THUMBWHEEL	.0099	.0209	.0491	.0158	.0957	.0957					.0412
13	1	2936171 LINK	.0024	.0047	.0127	.0043	.0241	.0241			.0241		
14	1	2936391 WASHER	.0062	.0001	.0002	.0013	.0078	.0078			.0078		
15	1	2197678 RETAINER	.0010	-	-	.0001	.0011	.0011			.0011		
16	1	6027529 DRIVE SCREW PROT	.0022	-	-	.0001	.0023	.0023					
17	1	2934158 HOUSING	.0697	.0161	.0436	.0204	.1498	.1498					
18	1	2936384 SHAFT	.0039	.0017	.0022	.0009	.0087	.0087				.0087	
19	1	2936385 RACK	.0249	.0105	.0285	.0110	.0749	.0749		.0744			
20	2	1022297 ASH NUT	.0009	-	-	.0001	.0010	.0020		.0020			
21	1	2936170 HOUSING	.0404	.0270	.0649	.0208	.1531	.1531					

Remarks: Mfg. Cost = Labor + Mtl + Burden; V — F = Fixed   FV = Full Variable — Purch. Fin'sh Cost (PFC) = Mfg. Cost + Burden + Other — Profit Improvement Target (%) = $\frac{C_I - V_I}{C_I} \times 100$

Cost Totals: 1.0000 | - | .1100 | - | .0967 | .0588
Value Totals: .35 | 0 | 0 | - | 0 | 0

Figure 12.12   Cost–Function–Value Analysis

a function basis, and how much each function costs. It is now necessary to determine if the functions are worth what they cost. The functions must now be evaluated and a value set for each function.

Evaluate the Functions — In this case the *modulate air* function costs $0.11 and the entire *modulate air system* $0.27, or 27% of the total product cost. This is the cost of functions 2, 4, and 5, which are required to modulate air. Is it worth the cost? In this case the car already had a four-speed blower to modulate air flow. Therefore, the team decided it was not worth $0.11, so the value set on *modulate air* was zero; however, it was necessary to shut off air flow. A review of specifications and

Figure 12.12 (continued)

competitive equipment practices showed that shutoff could be combined with vane rotation at a small additional cost. If the *modulate air* function is worthless, then all of the lower order functions must be considered worthless because they are only there to make the higher-order function work. Thus it was determined that the entire *modulate air* system cost of $0.27 could be eliminated. The result was a 27% saving per assembly. Other changes were made to bring the total saving up to 39% with very little expense.

Function–Cost–Value Analysis — When the function–cost–value worksheet has been completed, the differences between the individual function cost and values indicate the TFO. In this case, it can be seen that there are five functions that show a TFO greater than five cents each. These

FUNCTION	VERB	NOUN	TFO
2	Modulate	air	0.110
4	Direct	force	0.096
5	Apply	torque	0.059
7	Direct	air	0.062
12	Interlock	vanes	0.077
		TOTAL	$ 0.404

Figure 12.13 All Functions Over $0.05

Function	Function Cost Elements
Modulate Air	[f2 ($0.11) +$f$ 4 ($0.09) + f5($0.059) +f6($ 0.03)] = $0.20
Direct Air	[f7 ($0.21) +$f$ 14 ($0.05) +$f$15 ($0.02) +f 11 ($0.00) + f5 ($0.00)] = $ 0.28

Figure 12.14 System Cost

functions are shown in the chart in Figure 12.13 and form the basis for creative questions in the Creative Phase of the Job Plan.

Further examination of the Argus Chart shows that the target for opportunity is at least $0.40, as identified in Figure 12.13.

System Cost — *Modulate air* is a basic function for the *modulate air* system. When it was decided to modulate air the method decided upon to accomplish the function caused a whole system to be developed. This system is represented by the secondary functions: *vary opening, direct force, apply torque,* and *limit rotation.* Therefore, the cost of the *modulate air* system is the sum of the cost for each function, or $0.20, as shown in Figure 12.14. In addition, the cost of the direct air system would be the sum of all of the secondary functions added to support the system, or $0.28.

Argus Chart Application — Application of Argus Charts to far more complex assemblies and problems, although they require more sophisticated analysis, have been equally successful. However, all of the basic tools necessary to make a more complex Argus Chart have been illustrated here.

In addition to aiding in improving the value of existing designs or procedures, the Argus Chart can be used to develop new designs of products and as a tool to review management problems. It has been used as a guide to determine the most effective use of manpower, to review an interplant procedure and recordkeeping system, to develop the critical and supporting functions for a long-range plan, and as an aid in a marketing study to determine what to emphasize in sales literature.

The logic forces a decision. At first there may appear to be no satisfactory solution, but the clear and concise terminology required, in conjunction with the questions, digs out functions that may have been overlooked, taken for granted, or not thought of before.

The two-word definitions, explicit and within the framework of the Argus Chart, also become a highly potent sales tool in developing support for a new idea or proposal. The diagram provides a system analysis, pinpoints the problem, and helps to sell the change.

The combination of function thinking, Argus Chart, and creative problem solving provide a complete system for profit improvement from concept to production. This system can have an effect on your operations that will be limited only by your knowledge of the techniques and our willingness to apply them.

Identification of the TFO completes the Information Phase of the Job Plan and brings us up to the second phase, the Creative Phase. Completion of the Information Phase takes about half of the workshop time. It is also the most important in that it clearly identifies the problem and provides a firm basis for the development of solutions. Up to this point we have been applying and using our analytical powers, which in most cases have been well honed. The next step requires imagination and free thinking to develop new and unique solutions to the problem. Frequently this means we have to stretch our mental abilities to make greater use of our creative abilities.

Develop New and Unique Solutions — Step 2

In the Creative Phase we must work as uninhibited as possible to come up with solutions that may never have been thought of before. Creativity has been discussed in Chapter 11. In the example the first step was to brainstorm the functions identified as TFO. Each function was brainstormed following the rules in Chapter 11. An example of some of the ideas developed is Figure 12.15. In most cases hundreds of ideas will be developed and there will be some duplication. After screening out the duplication a group of potential opportunities will remain and these will be developed into concepts. This screening process brings us to the third or Evaluation Phase.

Screen All Ideas — Step 3

In many ways selecting the best ideas from a large group of data can be one of the most difficult tasks in a VE workshop. The first step is to quickly

2. **Creative Phase**   Sheet 4

**Develop
Ideas**   X/XX/XX Date

1.	bellows type flutter valve - one piece
2.	partitioned hollow sphere
3.	ball type with iris opening
4.	control flow by pinching off
5.	vane shut-off
6.	shut off by rotating louver
7.	living hinge vane - one piece molded interlock
8.	fluidic control
9.	double set of vanes movable vertical and horizontal
10.	two large vanes and louvers - duck bill type
11.	lower cost material
12.	replace thumbwheel and link with push pull
13.	chrome plate only exposed edge of thumbwheel
14.	vacuum metalize thumbwheel
15.	eliminate door shut-off
16.	shutter type shut-off torsion operated
17.	simple disc door in round opening rotated in door axis
18.	put outlet in top of instrument panel rotate 360 degrees (ship's funnel)
19.	elastic slotted diaphragm which varies opening size by push pull
20.	window shade type horizontal shut-off
21.	braided tube type like Chinese finger puzzle
22.	aircraft type vents
23.	ball with cleanser type valve
24.	rootes type ball
25.	one central outlet in place of three
26.	make housing of poly-pro incorporate door with living hinge push pull
27.	horizontal clamshell fixed vanes up and down
28.	plastic and metal pinched type door louver
29.	one piece molded fold-up living hinge vanes & housing - sonic welded
30.	
31.	
32.	
33.	
34.	
35.	

Figure 12.15   Screened Brainstorm List

look at the apparent overall cost to develop the rough concepts. This may cause the team to drop one or more of the alternatives as not profitable. The project life may not be long enough to pay for the cost to change, or the concept may not be compatible with the overall product assembly. Whatever the reason, elimination of several items will simplify the selection.

Worksheet No. 5, Figure 12.16, was designed to provide a quick look at alternatives from a cost basis. In the spot cooler example, Figure 12.16 shows the living hinge and hollow sphere design as the most costly and

	Evaluation Phase Screen All Ideas			Sheet 5 X/XX/XX Date

	Idea	Cost per pc.	Tool cost	Develop.	Total cost
1.	*lower cost material*				*-0.03/assy.*
2.	*Partitioned hollow*	*0.34*	*$ 40,000*	*$ 15,000*	*$ 55,000*
3.	*vane shut-off*	*0.63*	*$ 15,000*	*$ 8,000*	*$ 23,000*
4.	*replace thumbwheel*	*0.77*	*$ 3,500*	*$ 1,000*	*$ 4,500*
5.	*link w/wire*				
6.	*f/c spot cooler*	*0.67*	*$ 17,000*	*$ 10,000*	*$ 27,000*
7.	*remove chrome from*	*0.83*	------	------	-------
8.	*thumbwheel*				
9.	*same as present but*	*0.465*	*$ 45,000*	*$ 34,000*	*$ 79,000*
10.	*all living hinge*				
11.					
12.					
13.					
14.					
15.					
16.					
17.					
18.					
19.					
20.					
21.					
22.					
23.					
24.					
25.					
26.					
27.					
28.					
29.					

Figure 12.16 Evaluation Phase

the elimination of the *modulate air* feature as the next best alternative
from a cost standpoint. These alternatives were carried over to worksheet
No. 6, Figure 12.17, to look at the advantages and disadvantages of each.

From the evaluation of the results of these worksheets the team elected
to choose the elimination of the modulate air function and to incorporate
air shutoff into the vanes as their primary recommendation. This concept
seemed to offer maximum benefit with the least expense and the simplest
possibilities for retrofit into the existing design.

4. Evaluation Phase   Sheet 6

**Screen All
Ideas**   x/xx/xx Date

Idea *Partitioned Hollow Sphere*		Idea *Eliminate door - vane shutoff*	
Advantages	Disadvantages	Advantages	Disadvantages
Simple design	No independent modulation	Lower cost	Performance compromise
Few pieces		Less parts	
Good reliability	Requires design consideration to limit knee knocker problem	Less expensive material	Modulation not independent of direction
Low assembly		Minimum eng. design time req'd	
Minimum dup. cost	Styling limitation		
Consider in conj. w/new panel	Must be made in conjunction with new panel design	Fits existing assembly	
		Meets competition	
		Select for immediate application	

Living hinge design			
Advantages	Disadvantages	Advantages	Disadvantages
Low cost	High development cost		
Provides all functions			
Styling flexibility	Eng. experience limited		

Function Value

Best alternative from above $_____   Measure of Value = $\frac{Function Value}{Present Cost} X100 = ___$%

Present Cost $_____

Figure 12.17   Evaluation Phase

Develop the Recommendation — Step 4

Once the decision was made as to what the recommendation will be, the team progressed into the Planning Phase and began to develop as much detail as possible within the time available. Worksheets 7, 8, and 9 are designed to aid in forecasting any problems and roadblocks that might arise during development and implementation of recommendations. They are, in effect, a checklist for project success. Completing this series of worksheets is not just a "fill-in-the-blanks" exercise. The information developed during this phase of the project will provide the necessary information for the management presentation and to answer the questions and challenges that are sure to follow. Our experience has been that diligent completion of the worksheets will ensure that the team will be able to respond convincingly to any questions or challenges from the audience at the management presentation. The result will be that your recommendations will be considered credible and your potential for a successful result will be increased.

Sheet No. 7, Figure 12.18, identifies potential roadblocks, indicates where they may arise, and identifies the action required to eliminate them. Two concepts are considered to cover the event that a serious flaw in the favored concept may be discovered during the evaluation.

Worksheet 8, Figure 12.19, begins to consider the requirements necessary to sell the recommendation. This worksheet identifies the departments, supervisors, and managers who will be required to contribute to the implementation in some way whether by design and testing or simply by supporting the change. It indicates the action required, by whom it will be taken, and what the expense may be. It also identifies any problems that may arise.

Worksheet No. 9, Figure 12.20, considers the problems identified on Worksheet No. 8 and suggests means to eliminate each problem. In some cases it may be a good idea to have responsible people sign off on their responsibility. This is not so much to lock a person into the project as to maintain a record of approval. In large companies people tend to move from job to job over a period of time, and it is often advantageous to be able to show a new person that the project was accepted by previous managers.

To aid in developing useful information during this phase of the evaluation process, a Table of Excuses and one of Roadblocks is provided in Figures 12.21 and 12.22. The intention is to provide a checklist for the team to remind them of the fact that, in most cases, new ideas or changes are resisted and to provide a sample of the objections that may be offered to the recommendation.

Report to Management — Step 5

After all the screening and evaluation has been completed, the facts are all available to prepare recommendations for management action. The data developed up to this point should be available on the worksheets

Identify
Roadblocks   *X/XX/XX* Date

Select the best ideas	*Partitioned hollow sphere and/or all living hinge construction for rectangular units*	

Roadblock	Where / Why ?	Action required
High developments for living hinge.	*Engineering/no previous experience.*	*Start development early*
Sphere does not provide independent modulation.	*Engineering/does meet perf. objectives.*	*Review objectives and revise if possible.*
Styling/prod. planning	*Styling/limitation*	*Review early in program*

Alternative idea	*Convert current design to vane shut-off*	

Roadblock	Where / Why	Action required
No independent modulation	*Engineering/does not meet perf. objectives*	*Review objectives and revise if possible*
Effect on air direction unknown	*Engineering/perf. unknown*	*test*
Styling/prod. planning	*Styling/appearance*	*Review aesthetics*
Design change	*Prod. des./unscheduled*	*Review benefits and review schedule*

Figure 12.18   Planning Phase

and should be consolidated on Worksheet 11. This form identifies the project, indicates the state of the project before the workshop and basic recommendations for improvement. The advantages and disadvantages of the change are listed, as well as financial information. The financial information should include before-and-after costs, nonrecurring costs to implement the project, and a forecast of potential savings on a three-year basis. The intent of a three-year basis is that many projects may require a substantial expense to implement, thereby substantially reducing the net

4. **Planning Phase**   **Sheet 8**

How Do I
Sell It ?   *X/XX/XX* Date

Implementation document - Determine all data required. Attach additional sheets if necessary.

Persons and areas involved - List all departments and organizations that will be required to do something as a result of this proposal and the names and supervisors of those directly involved.

	Depart.	Supervisor	Action	Cost	
				Material	Labor
1.	4300	R. Kush	Product planning		
2.	6300	G. Neale	Styling approval		
3.	5720	H. Aluwe	Adv. Design development		
4.	3860	M. Daws	Revise perf.std. & testing		3500
5.	5210	T. Simpson	Approve matl. usage		
6.	3820	J. Kress	Incorporate in prod. design		1500
7.	2000	J. Jagge	Assembly		
8.	4300	R. Tenny	Purchasing concurrence		
9.					
10			tool cost	2500	

Potential problem areas - Indicate any forseeable problem area or unusual situation that might result during the implementation of the proposal

	Dept.	Problem
1.	3860	Vane shut-off may not prove directional control
2.	5720, 5210, 3860	Living hinge difficult to develop
3.	5720, 5210, 3860	Ball & socket require carefully developed specs
4.		
5.		
6.		
7.		
8.		

Comments: *Prepare product development appropriation request (PDAR) for submittal to management to obtain authorization & funds to develop recommendation - follow suggested format of attached sample on page IV.8. All necessary data is contained in worksheets -*

Figure 12.19   Planning Phase

benefit in the first year. However, in many cases once the initial expense has been paid, the project will be extremely profitable.

The presentation to management should be about seven minutes long and include about seven to 12 transparencies depending on the complexity of the recommendations. However, all presentations should include the following:

Title Sheet
  Name of project
  Value Engineering Workshop
  Date of presentation

**Project Study
Record**   x/xx/xx _Date_

Minimize Problem Areas — How might the problem areas identified above be minimized ?

	Dept.	Problem Area	How to minimize
1.	3860	Directional control	Test
2.	5720, 3860	Living hinge development	Buy temporary tools early
3.	5720, 3860	Ball & socket development	Buy temporary tools early
4.			
5.			
6.			
7.			
8.			
9.			
10.			
11.			
12.			
13.			
14.			
15.			

General verbal approval obtained — Preliminary discussions with people directly involved will usually pave the way usually pave the way for quick formal approval of the implementation document. Note the result of the discussions here.

Name	Comments	Approval	Date
Principle decision maker			
Others directly involved			

Figure 12.20   Planning Phase

Team Identification
    Name of members and area of expertise
    Design, engineering, manufacturing, purchasing, etc.
Modified Argus Chart
List of Recommendations
    Short term, three items
    Medium term, two items
    Long term, one item

TABLE OF EXCUSES

To save time for management and yourself, please give your excuses by number. Specialized excuses may be added for special circumstances, but this list should cover most situations.

1. That's the way we've always done it.

2. I didn't know you were in a hurry for it.

3. That's not in my department.

4. No one told me to go ahead.

5. I'm waiting for an O.K.

6. How did I know this was different?

7. That's his job, not mine.

8. Wait 'till the boss comes back and ask him.

9. I forgot.

10. I didn't think it was very important.

11. I'm so busy I just can't get around to it.

12. I thought I told you.

13. I wasn't hired to do that!

Figure 12.21 Excuses

Short-Term Recommendations
    Group, as practical
    Before and after
    Advantages
    Disadvantages
Medium- and Long-Term Recommendations
    Group separately and identify as above
Financial Profit Improvement Benefit
    Total piece cost saving for recommendations that can be implemented at one time
    Number units per year
    Gross saving
    Cost to change
    Net saving 1st year
    Net saving lifetime (at least 3 years)

ROADBLOCKS
Killer phrases that chloroform ideas and put men's minds to sleep.
(Checklist of 101 idea stoppers for negative thinkers.)

1. Our place is different.
2. That's beyond our responsibility.
3. That's not my job!
4. Not enough help.
5. Why?
6. It's against company policy.
7. It runs up our overhead.
8. We don't have the authority.
9. That's too ivory tower.
10 Let's get back to reality.
11. What do our competitors do?
12. Can't teach an old dog new tricks.
13 Good thought but impractical.
14. Let's give it more thought.
15. Top management would never go for it.
16. Let's put it in writing.
17. We'll be the laughing stock.
18. Not that again.
19. We'd lose money in the long run.
20. Where'd you dig that one up?
21. We did all right without it.
22. What can be expected from the staff?
23 It's never been tried before.
24. Has anyone else ever tried it?
25. What's the use?
26. Not enough time.
27. Too hard to sell.
28. I don't see the connection.
29. It won't work in our industry.
30. What you are really saying is...
31. It won't work in my department.
32. It won't stand up.
33. Let's all sleep on it.
34. You're right but...
35. I'm not convinced.
36. We've tried that too but...
37. We did it this way.
38. It won't work.
39. It's not in the budget.
40. Where does the money come from?
41. You can't do that!
42. You should know better.
43. We're not ready for that.
44. Everybody does it this way.
45. No, no, no.
46. Too academic in its approach.
47. It's not timely.
48. It's a gimmick.
49. It's not progressive
50. Not for us.
51. It's too hard to administer.
52. No good!
53. Plain stupid.
54. Screwy.
55. Impractical.
56. Idea is too radical.
57. It's too complicated.
58. It's unsound
59. It's not feasible.
60. It's too difficult
61. Too theoretical.
62. Impossible.
63. Production won't accept it.
64. Personnel aren't ready for this.
65. Engineering won't approve it.
66. The men won't go for it.
67. I can't sell it to management.
68. My boss won't like it.
69. Can't see it.
70. Too much trouble to get started-
71. We don't have the manpower.
72. Who is going to do it?
73. Takes too much time.
74. Too much work.
75. It's never been done before.
76. It will not apply to our problem.
77. Don't move too fast.
78. It's new.
79. It will set a precedent.
80. We have too many new projects now.
81. We don't want to do this now.
82. Not enough background.
83. Why can't we do this another way?
84. We have something just as good now.
85. Don't be ridiculous.
86. We know all this...
87. I'm too busy to decide now.
88. We haven't enough facts.
89. What about the directive?
90. It's not standard stock.
91. That will take two years to test.
92. It will make present equipment obsolete.
93. We don't have enough volume.
94. It's not permitted by specifications.
95. It's not in accordance with standard plans.
96. Let's shelve it for the time being.
97. Let's form a committee.
98. Cost doesn't matter...
99 Why change it - it works.
100. Could a vender supply this for less?
101. We can't help it - it's policy.

Figure 12.22   Roadblocks

Return on investment in months
Implementation target date
Recommendation for Action
What action is required
Who should perform the action
When should it be started and completed

Two points to keep in mind for a successful presentation are:

Tell them what they need to know.
Don't tell them anything they don't want to hear.

The fact of the matter is if they want to know, they will ask. Our experience has been that, following the seven minute presentation, questions and discussion usually goes on for at least an hour, and during this time all of the questions are always answered satisfactorily by the team.

Spot Cooler Example Summary

The data for the spot cooler recommendation is shown on worksheet 11, Figure 12.23. These data were consolidated on the financial profit improvement summary used in the Report To Management, Figure 12.24. A sketch showing the initial product and the product as implemented and installed in the vehicle is shown on Figure 12.25.

The project was eminently successful and was incorporated into the automobile within three months of the presentation to management. In addition, the Argus Chart was used as a road map to identify high-cost functions for a future new design. The resultant final design was a part of the next new design instrument panel and was used for several years in several car lines and is shown in Figure 12.26.

The Argus Chart shown in Figure 12.11 was used as a guide in developing the final design. In reviewing the sketches in Figures 12.25 and 12.26 and the Argus Chart, it can clearly be seen how the Argus Chart helped to create an understanding of the function cost and how it affected the final design. For example, the Argus Chart showed a cost of $0.16 to contain the assembly. This function is now contained in the instrument panel at no additional cost, since it was designed into the product. In addition, the air conditioning ducts are now connected directly to the panel structure rather than to the spot cooler. These changes eliminated considerable redundancy and cost. Additional savings resulted by simplifying

	Change		Date
	Proposal		XX/XX/XX

Product			

Part Assembly
Spotcooler assembly   A/C & Heater   Part   No.   2936148

Reference
All carlines

Description of Charge

Present	Proposed
24 Piece assembly Direction control accomplished by 5 vanes mechanically attached by a tie-bar assembled in a rotating housing. Modulation is accomplished with a thumbwheel, link, shaft and door.	18 Piece assembly Replace shutoff door, link, thumbwheel and attaching parts with vane shut-off .capability. Use lower cost ABS plastic.

Advantages
Fewer parts to assemble.
Less expensive material and less material.
Minimal engineering time required.
Meets competition.

Disadvantages
Performance compromised.
Modulation control is no longer independent of directional control.
Engineering will evaluate.

Cost Analysis

Implementation			Variable Cost per piece					Potential Saving	
	Material	Labor		Material	labor	Burden	Total	Benefits are based	
Eng.		1,500.00	Present	0.3427	0.2338	0.4235	1.00	on prompt approval	
Dvp/	1,500	2,000.00	Prop.	0.13	0.22	0.4	0.73	Series	Net Savings
Test	2,500.00		1. Pc. Cost saving				0.27	1st yr	60,000.00
Tooling			2. Est. Number units				250,000	2nd	67,500.00
Equip.			3. Gross saving (1 x 2)				67,500.00	3rd	67,500.00
			4. Total Implementation Costs				7,500.00		
			5. Net saving 1st year (3 - 4)				60,000.00		
Total	4,000.00	3,500.00	6. Payout - Months 4/3 x 12				1	Life	195,000.00
Total Material & labor		7,500.00							

Figure 12.23   Change Proposal

and revising the retention and interlocking of the vanes. The final design contains no fasteners; the entire product is snapped together and the spring-like arrangement on the top and bottom attaches the entire unit to the instrument panel.

POTENTIAL PROFIT IMPROVEMENT EFFECT

PIECE COST SAVING ⬚ *$ 0.27*

NUMBER OF UNITS/YR. *250,000*

GROSS SAVING *$ 67,500.00*

NET SAVING 1st YEAR ⬚ *$ 60,000.00*

NET SAVING LIFETIME ⬚ *$ 195,000.00*

PAYOUT *1* MONTHS, TARGET DATE *19XX*

Figure 12.24 Profit Improvement Effect

This example should clearly indicate how a knowledge of cost and function can guide a designer to a design to satisfy required functions at substantially lower cost. The product had already been reduced in cost by 75% before applying VE. There did not seem to be any way to reduce cost further, but the Argus Chart and the Function–Cost–Value analysis pointed out an additional specific 40% cost reduction potential without affecting installation in the major assembly. In addition, by using the Argus Chart for guidance a new design was developed reducing costs still further.

The example also illustrates that although it is possible to identify ways to reduce the cost of existing products, it is also possible to use the same techniques to produce new, more profitable products. Figure 12.27 shows the lifetime history of the product. If the VE methods illustrated here were applied in the earliest stage of the product design, it is likely that the result would have saved a considerable amount of money in both product and assembly costs.

In reviewing the history of the product it is important to recognize that when the *modulate air* function was eliminated it was possible to remove all of the parts that performed the function. Although the parts were removed, it was the elimination of the function that made it possible to remove the parts. Before the function analysis, it was deemed that it was not possible to obtain the assembly for less money. However, when the *modulate air* function was considered to be of no value, it was possible to eliminate all of the parts that performed that function. Identifying functions that are not required or that can be performed in new less costly

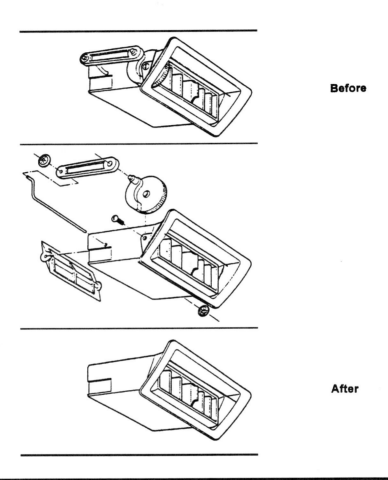

Before

After

Figure 12.25   Value Engineered Spot Cooler

ways makes possible the elimination of unnecessary cost in products and services of all kinds.

Implementation Phase — Step 6

An idea is not worth much if it is not put into practical use. Organizing for successful implementation is covered in Chapter 13, Organization and Implementation: Organization of the Basic Elements into a Practical System.

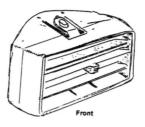

Front

Rear

Final Design Spotcooler
8 Separate pieces
81% cost reduction

Figure 12.26   Spot Cooler

	Original die cast	Cost reduced plastic	Value Eng'rd plastic	Target Value	Final Design plastic
Parts	24	24	18	--	8
Unit Cost	$3.80	$1.00	$0.65	$0.35	$0.19
Parts reduction	----	0	25%	----	66%
Cost reduction	----	73%	35%	65%	81%

Figure 12.27   Spot Cooler Product History

References

1. Miles, L. D., *Techniques of Value Analysis and Engineering,* McGraw Hill, Inc., New York, 1961.
2. Davidson, H. K., and Matthews, J. H., *History of the First Five Years of The Society,* Society Of American Value Engineers, 1962.
3. U.S. Army Management Engineering Training Activity, *Principles and Applications of Value Engineering,* 3rd ed., Aug. 1978, 3-36.

4. Fallon, C., *Value Analysis,* Triangle Press, Irving TX, 1978, 71.
5. Mudge, A. E., *Value Engineering, A Systematic Approach*, A. E. Mudge, 1981, 140.
6. Value Analysis Inc., *Value Analysis Engineering Manual,* Schenectady, NY, 1963, 3-1
7. Tsuchiya. H., *Value Engineering,* Matsushita Electric Industrial Company, Ltd., Japan, 49.
8. Fountain, R.E. Value standards in *SAVE Proceedings*, Kaplan, M., Ed., 1963, 60.
9. Mudge, A. E., *Value Engineering, A Systematic Approach,* A. E. Mudge, 1981, 63.
10. Bytheway, C. W., Basic function determination technique, in *SAVE Proceedings,* Vol 2, 1965, 38.

Chapter 13

Organization and Implementation: Develop a Practical System

Introduction

The objective of a VE study project is the successful incorporation of recommendations into the product or operation. However, the success of a project often starts back at the beginning. Each project must be thoroughly analyzed to determine its potential for benefit and the probability for implementation. This is as important as the knowledge and skill required to apply the system to attain successful results.

An excellent idea is worthless unless it can be implemented. If it isn't implemented no one will obtain the benefit. It must also be implemented in the manner intended. Unfortunately, there have been many cases on record where the idea could not be implemented because of the high cost of change. There are other cases where the recommendations were not properly understood and implementation resulted in increased cost. This often results in disillusionment and the feeling that VE doesn't work on our problems. Actually, in most cases the real problem was that the project wasn't properly diagnosed. It wasn't that VE doesn't work; it was inefficient preliminary analysis and preparation.

It does not seem reasonable to expend the effort and funds to make a value study without first having done the necessary work to ensure that the project is practical, that it can be implemented and that the necessary funds and manpower are available.

Selection of projects is a part of the entire VE implementation process. Many times management will assume that any and every project will prove to be profitable. This is not always the case. The project must be practical in relation to its effect on the organization. A brief outline of the SIVE Analysis, A simple project selection technique is Figure 13.1.

To aid in the selection of projects, development of people, implementation of projects, and all of the other aspects necessary to successfully achieve the stated objective, we have prepared some guidelines. They are guidelines, not rules because every organization is different and successful VE operations must be integrated into operations to become part of the day-to-day decision-making process of the company.

To begin with, we will look at the overall organization and implementation of VE operations. Then we will look at some of the details that make for success.

Goal for Achievement

What do we want to get from VE? What will be the objective? This is the first question to answer. VE can increase productivity, reduce product cost, improve quality, reduce administrative costs, and provide a number of other benefits that may be critical to operations. Whatever the goal may be, it should be defined in specific terms: increase productivity by a specific percent, reduce product cost by a specific number of dollars per unit, etc. This goal may change as experience is gained in application of the methods. My experience has been that as greater understanding is gained, the scope of applications will broaden from products to services in manufacturing and administration and other areas of an organization. Whatever the initial goal may be, it can be revised and broadened as skill in application and implementation of the process develop, and as understanding and credibility increase.

VE is a people-oriented process designed to help people do a better job by helping them break down constraints to understanding. It provides some very specific methods and systems to achieve results.

Some people perform a wide variety of jobs in an organization. It is certainly logical to expect that if they can be provided with a system that can help them do a better job, anything that they are expected to do can be improved. In the end, it is people who do the thinking. If we can improve their performance, everyone will benefit.

PROJECT SELECTION

1. **Develop Awareness**
   Products
   Operations
   Investments

2.. **Selection Methods**
   Intuitive
   Scientific

3. **Considerations**
   Noncompetitive product
   Low Volume
   High warranty
   Quality problems
   Supplier problems
   Manufacturing difficulties
   Capital investment
   High manpower requirement
   Bottlenecks
   Potential market
   Government regulations

4. **SIVE Analysis**
   List potential projects

Potential saving	S	
Implementation cost	C	$R = \dfrac{S}{C} \times F$
Confidence factorF		
Project priority	R	

   Confidence factor - F
   Poor   1
   Questionable   2
   Fair   3
   Good   4
   Very good   5

5. **Example**

S = $60,000.00	$20,000.00	$2,000.00
C = $10,000.00	$10,000.00	$ 500.00
F = 1	5	4
R = 6	10	16

Figure 13.1 SIVE Analysis: Scientific Investigation of VE Project

This has been our experience. Many people, highly skilled in their jobs have developed new insights that have created breakthroughs in technology as well as major organizational and operational improvements.

The goal for achievement should be known to everyone. It can be product oriented or directed toward manufacturing or administrative operations. It need not be company wide. However, the scope can be broadened at any time. Once the goal has been determined, the means to achieve the objective can be developed.

Develop a Plan

There are five steps to incorporating VE into operations. They are:

- Evaluate the system
- Define an objective
- Develop a plan and organization to achieve the objective
- Understand the principles
- Implement the plan

Each step can be approached in a number of ways. However, there are certain specific problems to be considered and pitfalls to be avoided in each. Understanding the problems and pitfalls rather than outlining a specific method or procedure should provide the necessary guidelines for an effective operation. In many cases a consultant can aid in the initial stages and support each step of the process by providing a broad range of experience for the client to build upon. However, it is important that the consultant have the type and quality of experience to ensure success.

Evaluation of the System

Evaluation of the VE system can be a very tricky process. There are companies who have spent very large sums of money for educational seminars and workshops and are not using the systems in any way. There are companies who did not understand the principles, and when they tried to apply them found that they had neither the skill nor the discipline to achieve success. There are others who feel a highly organized cost reduction program is VE. As a result there are some who feel that VE works but not on their product. There are others who feel that VE is nothing new; it's the same old thing they've done for years under a different name.

Evaluation of the benefits to be obtained from VE should, therefore, be based on at least some prior knowledge of the methods and disciplines so intelligent questions can be asked to determine what is being done. Are the principles of function analysis and evaluation being applied? Is the Argus System of function analysis or some other FAST system being used? How is the creative stage handled? How are projects selected and organized? How is the team approach used? What authority does a VE team have to implement projects? How are teams selected? How is the operation organized? These are key questions that are required to evaluate whether a company actually has been using VE based on the principles established by Lawrence Miles and supported by the Society of American Value Engineers (SAVE).

A major element of the evaluation process should be a one-day orientation for key management, those who will be required to support operations with time, manpower, and funds. The orientation should be presented by one who has had successful experience conducting VE operations within the constraints and limitations of daily operations. Preferably the person should be certified as a Certified Value Specialist (CVS) by SAVE.

Just to understand the principles is often not enough. How to make them work in an operating environment is frequently at least of equal importance. As in everything, future success is based on a firm foundation.

Understanding the Principles

Very early in the plan to introduce VE into operations high-level and operating management must be introduced to the system. The intention is not to teach them VE but to demonstrate the benefits to be achieved and how they are produced. This establishes the need to apply the process and defines the necessary commitments for success. Those who should attend would be everyone who will be expected to support operations with time, manpower, and funds.

It is difficult for a large group of high-level people to attend a one-day seminar. However, it is essential for successful operations. The method sounds so reasonable that the first reaction in many people is that we do that all the time. Participating in a one-day seminar offers the opportunity to actually practice the various techniques and enables the participant to realize both the mental difficulty and benefit of the system. Attendance also broadcasts the message of importance to all levels of an organization. In addition, the managers attending often derive substantial benefit from the session which can achieve immediate results.

The one-day orientation should be a case study so participants can try the various methods and systems. The result will be an understanding of the system and how it may be applied to various projects. It will identify the organizational and operational pitfalls and, in many cases, define projects for future workshops.

Completion of the management orientation will create a need for a decision to determine how operations will proceed from this point. In most cases several people will volunteer a project to evaluate or at least indicate they are interested in discussing one.

If a consultant has been brought in, he or she will now be able to assist in developing effective continuing operations. If one hasn't been brought in, now would be a good time. The consultant's experience can ensure success from the start of operations.

At this point there are two ways to go. However, in the long run, the same objective will be achieved. One approach is to select five or six projects for analysis and conduct a multiproject workshop directed toward indoctrinating a large group of people (30 to 35 per workshop) in the system and at the same time demonstrating the process by achieving concrete benefit for the organization. An outline schedule for a SAVE-approved 40-hour workshop is in the Appendix. This workshop contains all of the essential elements for a basic VE educational program and has been expanded to be a graduate-level college course at several universities.

Participants in this workshop will learn the process by applying the methods to projects of current interest to the company. Participants should be relatively high-level people who will be expected to support operations in the future. In addition to education, these workshops usually develop substantial monetary benefit for the company.

The second approach is one or two teams working on a specific project of current interest to the company.

Both methods can be successful. However, the first is better suited to very large organizations with large amounts of manpower. The second can be used in both large and small organizations and produces substantial benefit that can be used to aid further development.

In many cases a combination of the two, plus a series of orientations, can be used effectively. The specific plan depends entirely on the organization and should be tailored to fit. The educational pyramid in Figure 13.2 illustrates the scope and extent of educational programs in any organization.

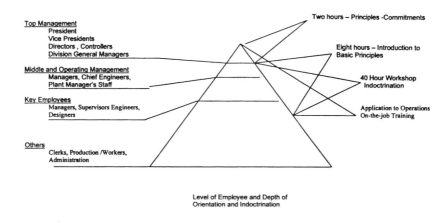

Figure 13.2 Education Pyramid

Organization

The first step is to determine the objective, which was discussed earlier. The second should be to develop a plan to achieve the objective and set up the necessary organization. The third step is implementation of the plan; the fourth step is follow-up and audit operations.

The essential elements are:

1. Define the objective
2. Develop the plan
3. Implement the plan
4. Follow-up and audit operations

Upon completion of the evaluation and with a decision made to implement VE the first step should be to appoint a VE Coordinator. A brief outline of factors to be considered in selecting a VE coordinator or manager is Figure 13.3. The coordinator will develop and organize a plan for management approval. Inherent in the plan should be education and application programs for all who will be involved in operations.

The coordinator should be required to select a consultant, develop an educational plan, aid in organizing and conducting workshops, and identify people who may be developed into value specialists. The extent of these programs will depend upon the size and scope of company operations as well as the urgency of results.

FACTORS TO BE CONSIDERED FOR A
VALUE ENGINEERING MANAGER

Primary Purpose of Position

Establish the Value Engineering business discipline as a part of the fiber and decision making process of the company to increase the opportunity to maximize the profitability of all products marketed by the company.

Plan, staff and direct a Value Engineering program to provide maximum product value by the application of recognized techniques to identify and eliminate unnecessary cost in the products and operations.

Develop and implement a program to educate employees, management and suppliers in the Value Engineering approach to problem solving with particular emphasis on function and value.

Publicize and demonstrate the use of Value Engineering techniques to company management and suppliers to develop support and participation in the use of Value Engineering and in the implementation of recommendations.

Knowledge and Skill Requirements

Degree in engineering, business or economics with a thorough understanding of technical aspects of product design and development, business operations and economic factors involved. Value Engineering workshop training.

Must have three or more years in Value Engineering program operations and a thorough understanding of the techniques and methodology as applied to both product development and manufacturing operations.

Minimum of ten years combined experience in product management, project engineering, manufacturing management or product development with a thorough understanding of procurement practices, systems analysis, cost estimates or any of a number of other broad rather than specialized product areas.

Must be creative, flexible in planning and thinking and have demonstrated leadership abilities necessary to organize and guide persons of widely divergent backgrounds into an effective team.

Must be effective communicator in both oral and visual techniques.

Figure 13.3   Value Engineering Manager

From what we have noted here it is apparent that the problem is complex from the standpoint of options. However, successful operations do not have to be extensive. Starting small and developing successfully is preferred to a lot of noise and a big crash because of poor planning. It is best to plan for the long term.

Attitude — One of the most important factors in VE is attitude on the part of management and people on task teams. A positive, cooperative, supportive attitude is required. In many cases VE requires a new management style. It cuts across organizational lines, looks at taboo aspects of a problem, and recommends drastic changes compared to the past. To accept these disruptions to the old way of doing business requires faith and understanding — a positive attitude.

In most cases whenever a new idea is presented to an American group the initial reaction is negative. The first remarks are, "It is interesting but let me tell you what's wrong with it." The best approach to this reaction is to listen carefully. They may have some ideas you overlooked. However, after all of the negative reaction has run out, be prepared to ask some specific positive questions of the person or group that will elicit positive responses. For example, "I understand your difficulty in producing this part in the plant. What do you think we would have to do to make it practical? Do you see any changes we could make that might satisfy your methods?" This will usually develop a positive result.

Never argue. In many cases it is beneficial to solicit negative ideas, but be prepared to develop positive questions. Our attitude is that we must begin to ask, "What's good about this idea?" "How will it help us do a better job?"

Changing people's attitudes is a difficult job and may never happen, but understanding the reasons behind the negative reaction should make it possible to persuade most people that they can benefit from success. Remember, there is a risk of failure in new ideas. New ideas require change, and the change may not work. They want proof it will work before they will support it. However, maybe you can show the benefits are worth the risk. Remember, there can be no progress without change.

A good way to change people's attitudes is to show that top management is interested in VE and expects participation and results in achieving stated goals.

Value Council — The value council is a small group of high-level executives who oversee operations. In a small company it might be chaired by the president or in a large company by a division manager.

The council should be staffed with people who have the authority to make decisions relative to acceptance or rejection of proposals, authorizing funds and manpower. They set the attitude, develop the environment, break bottlenecks, and by their interest and visibility, create credibility and participation and provide authority to operations.

It is important that members of the council make every effort to attend council meetings except in cases of dire emergency. When a member is unable to attend he should authorize his key assistant to act for him. If the council attendance degenerates, the message is, "We are losing interest."

PROJECT STATUS - VALUE ENGINEERING WORKSHOPS										PROJECT STATUS 7-1-X4			
	MIN. POTENTIAL SAVING EST. AT WORKSHOP (THOUSANDS)	ESTIMATED SAVINGS – RELEASED TO PRODUCTION OR IMPLEMENTED (THOUSANDS $)											
PROJECTS	1st YR NET	19X1	19X2	19X3	19X4	19X5	19X6	10X7	DROPPED	DORMANT	ACTIVE	RELEASED
Workshop #1												
1 FUEL VAPOR SAVER	1,060	1,336	1,636	1,536								X
2 PARK. BREAK MECH.	100								X			
3 R.R.WIND . DEFOGGER	148					118	118	118				X
4 WDO REGULATOR SYSTEM	309	110	110	110								X
5 HOOD LACH	143					207	350	360				X
6 A/C OUTLET	150	136	136	136								X
TOTAL	1,910	1,681	1,781	1,781		326	468	468				
WORKSHOP #2												
1 HEATER & A/C CONTROL	1,721		1,461	1,781	421							X
2 REMOTE MTG. TRN. SIG.	200		72	150	150							X
3 SEAT TRACK	300								X			
4 REMOTE O. S. MIRROR	117								X			
5 STEERING COLOMN	1,742								X			
6 SPEED SHIFTER	153								X			
7 ACCEL. CONT. LINK	146										X	
TOTAL	4,479		1,623	1,931	671							
1												
2												
3												
4												
5												
6												
7												

Figure 13.4   Project Status Audit

The council should be made up of five to six people. Their duties are:

• Set objectives
• Guide operations
• Monitor progress
• Eliminate roadblocks
• Recommend/approve projects

Audit Results — There are two reasons to audit results. The first is to determine the actual benefits received. Is it in accordance with expectations? If not, why not? The second is to determine how to improve operations.

A periodic status report on a project tends to move it along. This is especially true of cost reductions. A typical audit sheet is shown in Figure 13.4.

IMPLEMENTED SAVINGS (THOUSANDS $)			COMMENTS	CONTACT	VALU ENGINEERING - IMPLEMENTATION PHASE
RELEASED	EST. TO DATE	Auth.			
4,408	4,408	C101	REFLECTS MINI AND MICRO MINI EXPANSIONS TANKS		
		C129	NEW CONCEPT OVERSHADOWED PROPOSAL		
354		C694	COMPLETE A - BODY		
330	330	C252	COMPLETE		
307		C314	COMPLETE - IN EFFECT ALL MODELS (INCLUDES 4-3)		
406	406	C147	IMPLIMENTED		
6,404	5,143				
3,653	3,653	C237	IMPLEMENTED		
372	372	C487	IMPLEMENTED		
		C173	POOR PAYOUT		
			WORKING ON 3 WIRE CONTROL - L.H. SIDE		
			REQUIRES ADDED INSULATION - NO COST SAVING		
		C115	NO MARKETING REQUIRMENTS		
		C219	NO ACTIVITY		
4,026	4,025				

Figure 13.4   (continued)

Concluding Comments

This is a very brief outline of some of the factors to be considered to implement VE in your organization. The complete subject would require many volumes, but even then there would be many exceptions.

VE is a task-type system. Set up the group get the job done, dissolve the group, get on with the next project. It is people oriented; it is designed to get maximum performance from the individual and capitalize on that performance by supplementing it with the group. Of course there must be some type of staff to provide continuity, and they must be skillful in application or know how to get people who can produce results.

Remember, success is based on the three A's: Awareness, Attitude, and Application. There must be a positive attitude in the origination, an awareness of the need to change, and the skills to apply systems for effective results.

If these guidelines are followed it has been proven that the benefits will be almost immediate and far greater than the usually expected results. They are often outstanding.

Chapter 14

Examples: Winning Results

Introduction

How to solve your problems or improve your operation has been described in detail in the previous chapters. However, in order to show the scope of potential application of the methods and systems discussed herein, the following examples should illustrate the potential benefit available to everyone.

It is not always possible to give complete details on projects because of legal requirements involving restricted information and other competitive limitations. However, we have provided as much information as possible to enable you to understand the potential for improvement from the application of function analysis and associated techniques such as the Argus System to a wide variety of projects. If additional information is required, we will make every effort to connect you to someone in various companies, if at all possible, who will be willing to discuss the projects with you in greater detail.

800-Ton Stamping Press

We had conducted a series of successful projects on various small presses ranging from 60 tons to several hundred tons for a company, a leader in the field. The success of these projects led the company to see what could

Figure 14.1   800-Ton Stamping Press

be done to increase the profitability of larger presses. These presses were usually built one at a time, and each incorporated a considerable amount of custom detail required by the customer.

It was desired to reduce the cost of an 800-ton stamping press by about 25% and we were asked to set up a program. The first task was to analyze product costs to determine the projects required to achieve the goal. The cost analysis indicated that 89% of the total variable cost of the machine was in the three product areas identified below. The complete machine is shown in Figure 14.1. The Argus Chart for workshop A is shown in Figure 14.2.

   A.   Crown Assembly
   B.   Slides and Uprights
   C.   Assembly Operations

The usual manufacturing procedure was to design and construct the machine in the plant and to thoroughly test the product before disassembling for shipment to the customer and reassembly in his plant.

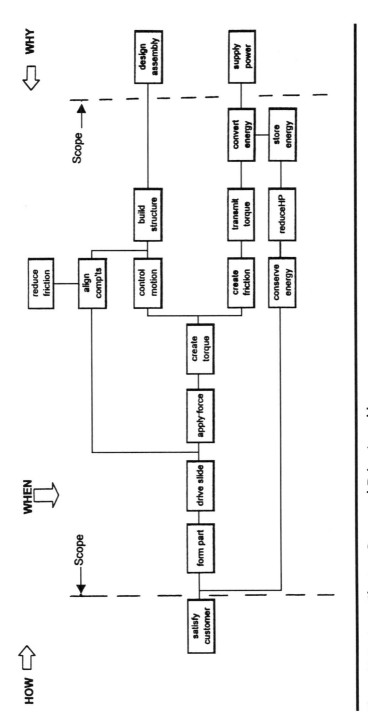

Figure 14.2 Argus Chart — Crown and Drive Assembly

Successful projects were conducted on the crown assembly and the slides and uprights. The crown was a self-contained unit and included the drive system. However, the slides were a part of the vertical structure, and any changes in the slide mechanism would probably lead to accommodating changes in the upright structure. For this reason the slides and uprights were considered to be a single project. The third greatest cost area of the project was the assembly operation and involved electrical, pneumatic, and hydraulic systems and considerable wiring and plumbing, as well as other labor-intensive work.

The variable or manufacturing cost of the product was about $350,000. The result of the three workshops was a $60,500 potential cost reduction for a new product cost of $289,500. This was a 17% potential total cost reduction that made the company highly competitive.

The workshop produced a series of specific recommendations, many of which were incorporated directly into the machine currently under construction and to other machines of different sizes.

Recommendations regarding assembly operations included the following:

- Improved painting methods and materials, airless spray
- Revised assembly methods to eliminate retapping
- Improved workforce utilization
- Relocation of electrical and tool cribs for improved material distribution
- Material handling and storage
- Laser instrumentation to measure and control tolerance stack-up
- Simplified product identification methods
- Pre-machine level block area

In addition, a number of product improvements were also cited, such as:

- Structural redesign and improvement
- Simplified drive train
- Main motor size
- Relocation of various components
- Counter-balance–purchased assembly

Benefits resulting from these recommendations included:

- Reduced cost
- Reduced number of components and material
- Reduced manufacturing labor
- Reduced assembly labor through improved utilization

- Reduced assembly time through improved material flow and manufacturing techniques
- Improved subassembly manufacture
- Improved scheduling processes
- Overall increased cost awareness
- Improved product quality

Automobile Dealership — Construction Schedule

At one time I was employed by an automobile company that had a very large realty company. The company built and maintained hundreds of dealer facilities across the country in addition to a number of other interests in housing developments, shopping centers, office buildings, and other commercial enterprises. The company had several hundred million dollars invested in dealership operations, and I was asked to see if I could help them reduce the length of time required for project development from the time the project was authorized to dealer occupancy.

At the start of the project the average total elapsed time from site selection and land purchase to construction and leasing was estimated to be 502 days. This was about 1½ years. History showed that there was a direct correlation between the availability of a dealer facility and the number of cars sold. This meant that a specific cost could be related to lost sales.

A review of the project process flow chart identified where the 502 days were expended in a project. This time included selecting and obtaining the option on the land, topographical surveys, soil borings, facility layouts, bid estimates and analysis, budget reviews, design, construction, and a myriad of other activities up to the time of dealer occupancy.

The process activities were identified, the functions defined and the Argus Chart, shown in Figure 14.3, constructed and evaluated. The 502 days of elapsed time included 30 functions. Three functions, resolve restrictions, obtain data, and construct facility, took 85% of the time. This evaluation was obtained by using time as a measure to evaluate the Argus Chart in the same way as money is usually used.

As a result of project recommendations, the project process procedure was revised to make it possible to conduct several of the long-term activities in parallel with other activities. For example, approvals for early site work were obtained from property owners before ownership was transferred so topographic surveys and soil borings could be made as soon as possible. In order to facilitate the process a standard plat was devised for soil borings that could be used in every facility that would

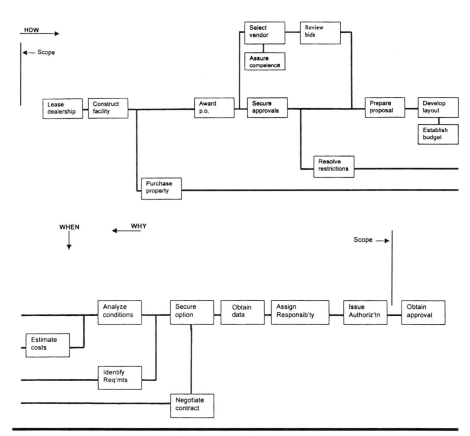

Figure 14.3   Argus Chart — Dealership Construction Schedule

assure the purchaser that the land met basic requirements. Standard designs were developed for several parts of the facility to reduce overall design and development time, and a single source contracting procedure was developed to reduce contractor project interface time.

The result of these recommendations was a potential average time saving of 262 days, or a 47% average saving in time per project. Based on the average annual construction program the yearly benefit in increased rents would be over $1,250,000 per year. The additional benefit from increased vehicle sales was not included in the benefit.

As a result of this project, additional projects were conducted to improve dealership operations, such as insurance claims adjustments and other dealership landlord responsibilities. In addition, a project was set up to determine an ideal dealership design for location anywhere in the country. An Argus Chart was developed, and the functions ranked and weighted. A competition was then set up with three noted architects to determine an ideal design concept. Only one of the architects chose to use the function requirements and incorporated most of the required functions in his concept and won the competition with an exceptionally creative design.

Automatic Clutch Adjuster

Most new cars with manual transmissions have an automatic clutch adjusting mechanism of some type to eliminate the need to periodically adjust the clutch. In many cases the engineer, when given an assignment, will look to see what had been done before to solve a similar problem. This frequently eliminates the need to "reinvent the wheel," but it may also carry over problems or practices from the earlier project. Such was the case in this clutch adjuster project. Something was needed in a hurry, so the first action was to see how others were doing the job. The result was as shown on the original project worksheet in Figure 14.4 and is basically a copy of other adjusters already on the market.

In this case we were able to bring the project to a VE workshop before the final design was completed. The result of the workshop was an entirely new concept, as shown in Figure 14.5 that offered a cost saving of 35% over the original design and a reduction of 53% in the number of parts. This simplification also contributed to quality and reliability. The actual production components are shown in Figure 14.6.

This design is simple and impressive because it required the breaking down of preconceived ideas to develop an entirely new mechanism. This process was aided by the use of function analysis and creative effort on the part of the team to develop different ways to perform the required functions and then to select the best way to satisfy the overall requirements. The process was aided by the construction of an Argus Chart and the function–cost–value matrix. Both of these tools opened the mind to the specific purpose of each component in the design and pointed directly to the high-cost, low-value functions. The high-cost, low-value functions then became the Targets for Opportunity (TFO) and the basis for creativity, in this case a brainstorming session, to develop ideas for evaluation.

This design was used in production vehicles for the entire extent of the production run of 5 years. Although this example produced a 40% saving in cost, the final design shows too many small parts and would have benefited from a Design for Manufacturing and Assembly (DFMA) analysis to limit them to only those absolutely required. In most cases these parts are added in late design stages to compensate for some

Information Phase   Sheet 1

Project
Information   _____ Date

Part Name		Drawing or part no	
Automatic clutch free play adjuster			
Used on (Name or Number)		Number required per assembly	
Team number 3		Workshop number and date	Task force and date

Team Members	Department	Phone number
John Shep	Rubber and Plastics	6-4570
Keith Moy	Transmission Design	6-3334
Mike Dever	Mech. Components Design	6-4448
Jim Duggan	Accounting and Systems	6-5589

Present cost

Total cost	Cost Elements		
$1.2477	Material $ 0.6728	Labor $0.1894	Burden $0.377

Estimated annual production   270,000 units

Operation and Performance

The automatic clutch free play adjuster maintains travel and force to operate clutch pressure plates during gear changes / shifting gears.
Performance is by means of a clutch cable that transfers actuation from clutch pedal to transmission. Clutch pedal has ratcheting device to maintain the travel and force during wear of clutch disc without the need for assembly or customer adjustment.

Includes other costs - See sheet 3

Figure 14.4   Original Design Project Worksheet

shortcoming of the assembly. At this stage of development, time is becoming short and the goal is to produce a working part. Unfortunately, the cost for these parts becomes buried in the final total cost and is rarely identified for future action. It is, therefore, imperative that careful scrutiny be maintained to be sure the assembly is constructed as closely to the original design concept as possible or that the design staff is made aware

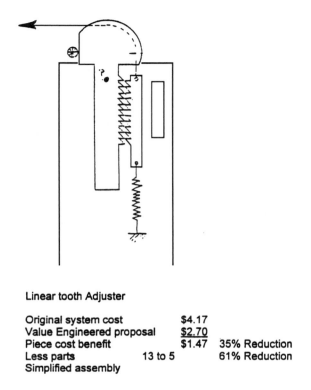

Linear tooth Adjuster

Original system cost	$4.17	
Value Engineered proposal	$2.70	
Piece cost benefit	$1.47	35% Reduction
Less parts	13 to 5	61% Reduction
Simplified assembly		

Figure 14.5   Value Engineered Design: Automatic Clutch Adjuster

of the consequences of adding seemingly minor items to the design without careful analysis.

Automotive Suspension Crossmember

It is always difficult to show examples of critical items because of competitive circumstances. In this case the project has been out of production for many years, so a competitive condition no longer exists. However, we are constantly being faced with similar projects in different companies that involve the very same problems faced in this project, so the project offers a current example.

The project involves an automotive front suspension crossmember. A sketch of the product is shown in Figure 14.7. The product was very heavy and costly and created serious warranty problems. One of the major reasons for the warranty problems was that the product involved 240 inches of arc welding that didn't always meet standards.

Figure 14.6   Actual Production Product: Automatic Clutch Adjuster

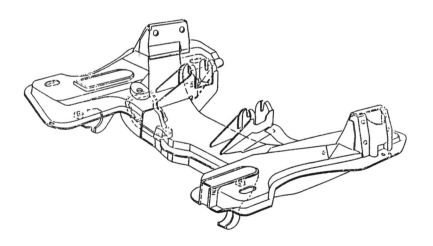

Figure 14.7   Automotive Front Suspension Crossmember

The project was brought to us by the head of the Materials Department. He had been charged with improving the product and, after careful analysis, was informed by his staff that the part was too complicated to develop a satisfactory modification. As a result, he asked if we thought we could offer any assistance. We made a diagnosis of the problem and agreed that we could help if we could get the required top-level people from design and manufacturing. We made a list of the people we wanted, and he agreed to get them assigned to the project.

The project was conducted over a two-week period; we worked two days the first week and three days the second. The Argus Chart developed by the team is shown in Figure 14.8. It has been simplified for illustration but includes all of the critical functions. During construction of the Argus Chart and analysis of the result the team became aware that forces in the design created serious torsional stresses. A redesigned structure realigned these forces to eliminate the torsional problem and resulted in a simpler, lighter, stronger frame for a substantially lower cost. The cross-section of the frame in Figure 14.9 shows the original and redesigned structures.

The basic change was to move the flanges down and realign the beam to align the forces and increase beam strength. These changes were made within the envelope of the current design to ensure compatibility with all mechanical relationships within the existing overall package. It was necessary to move the sway bar forward a few inches and add an indentation in the engine oil pan to ensure adequate clearances; however, these were minor changes.

Although the new design was a substantial change, it was compatible with existing manufacturing methods and equipment and offered a number of benefits. The new design reduced weight by 16 pounds and cost by 33% and made it possible to recover expenses in 5 months. This financial benefit was without taking credit for the reduced weight and cost of steel and additional weight reduction credit for performance. In addition, welding was reduced by over one third, and manufacturing stamping and assembly were simplified. Over the production run the overall financial benefit was estimated to be in excess of 40 million dollars.

Engineering Department Organization Analysis

The company was having budget problems. Business was down, and the cash flow was rapidly becoming a major problem. Top management kept pressing to reduce the Engineering Department budget, which was about 200 million dollars per year and employed 4,000 people, including engineers, technicians, designers, draftsmen, mechanics, specialists, and financial and supporting personnel of all types. These people were developing a complex technical project from concept through release to manufacturing.

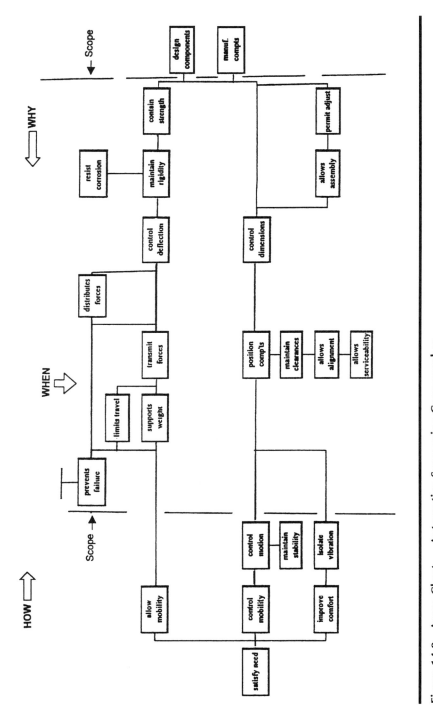

Figure 14.8   Argus Chart — Automotive Suspension Crossmember

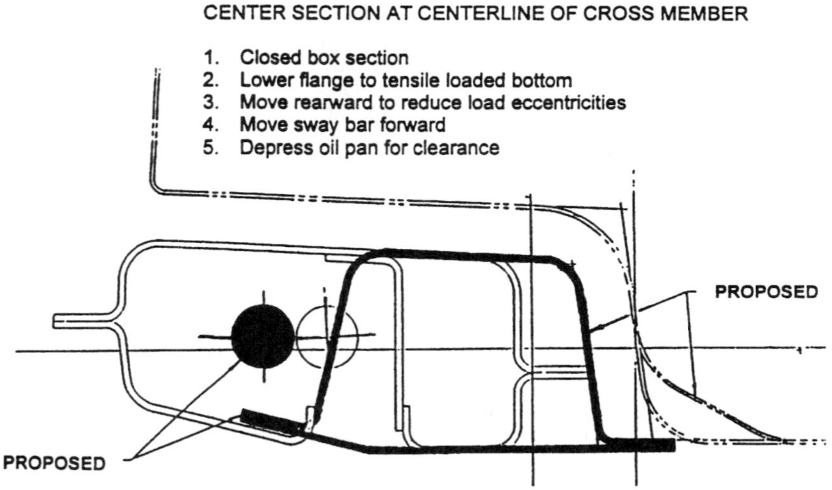

CENTER SECTION AT CENTERLINE OF CROSS MEMBER

1. Closed box section
2. Lower flange to tensile loaded bottom
3. Move rearward to reduce load eccentricities
4. Move sway bar forward
5. Depress oil pan for clearance

PROPOSED

PROPOSED

Figure 14.9   Recommended Suspension Section

The product included complex systems; a power plant, climate control, mechanical mechanisms, electrical and electronic controls, and a number of other systems. It was sold in a highly competitive worldwide retail market. The product was an automobile.

In the development process it had been assumed that there was a phase of operations that is not directly related to new product development, only to improving existing products in accordance with various requirements such as field service, competitive pressures, and others. This phase of operations was called continuing engineering. In times of financial stress it is important to know what this level of effort is so budget adjustments can be made to meet financial restrictions without having a detrimental effect on overall operations.

It has been our experience that when it is required to reduce the budget, no one wants to participate voluntarily. Everyone feels that they are already short handed and to reduce their staff further will prevent them from attaining their assigned mission. As a result, nothing gets done until someone in higher authority issues a directive to reduce every budget by 10 or 15% by "Monday" morning. The result of this directive is usually very serious damage to some operations, while others simply slide by. This reluctance is usually caused by budget procedures and makes a strong point for Zero Based Budgets in some form. The truth of the matter is the fact that many people are very busy performing tasks that are not overly important, while other more important tasks go undone.

verb	noun	verb	noun
create	design	prepare	plan
transmit	information	negotiate	alternatives
evaluate	information	evaluate	capability
confirm	design	allocate	resources
model	concept	appropriate	funds
Identify	deviations	support	operations
establish	criteria	authorize	program
utilize	resources	reconcile	differences

Figure 14.10   Partial List of Functions — Engineering Operations

CREATE DESIGN
The time required to generate a new system, assembly or component Includes coming up with the idea by anyone, design and layout time, engineer's calculations, programming time for the computer , etc.

Figure 14.11   Example of Glossary Item

In our case the situation was critical. The budget had already been cut several times, and no one knew where to look next. Since the objective of VE is to identify and remove hidden, unnecessary cost, we were asked to look into the problem to see if VE methods could be applied to achieve a benefit.

The first step was to organize a team. The team consisted of a chief engineer from each of the three major design–development operations and administration. In addition, the VE department financial specialist and two certified value specialists made up the team of six people. Because of the critical state of affairs, it was not possible to take the chief engineers away from their jobs for more than a few hours at a time so it was agreed to meet twice each week from 3 to 7 PM.

In our early discussions we very quickly found that the functions to perform continuing engineering were the same as for new product development. The only difference was the project task. With this understanding we listed the activities for the entire engineering operation, and the functions of each step were defined. An Argus Chart or FAST diagram was then constructed. After developing the Argus Chart considerable additional discussion ensued to clarify the functions as originally defined. The completed chart took 72 hours to develop and contained 72 functions. The chart was then thoroughly discussed to ensure that it covered all aspects of the operation, and a glossary of the functions was made to ensure future understanding. A partial list of functions appears in Figure 14.10, and an example of a typical glossary item is shown in Figure 14.11.

	No. Depts.	Manpower	Type Of Department Unique	Design	Design & Development	Test &Development	Admin.
Sample	18	1,097	5	2	4	6	1
Bal. Of Eng.	74	2,396	25	5	12	22	10
Total Eng	92	3,493	30	7	16	28	11
% of Total	20	31	17	29	25	21	9

Figure 14.12   Sample Department Classification

The diagram provided some interesting information. Most important, it showed that many functions were performed to satisfy functions outside the scope of engineering responsibilities. Many of these functions contributed to higher-order functions to support other company operations, such as the purchasing and legal departments, and in some cases to create an image in the community by collecting pledges to support charitable community organizations. The next step was to determine how much the functions cost and how the funds were distributed.

In the second step it was necessary to determine function cost and function value. To do this, a sample group of departments was determined by random selection processes, and the department managers were asked to distribute their departmental cost by function. The departments selected for the function–cost–value analysis are shown in Figure 14.12.

To do this, a Function–Cost–Value chart, Worksheet No. 3, was completed by each manager. A copy of one of the documents is shown in Figure 14.13. Every department manager was familiar with the VE system and knew how to convert activity cost to function cost. The glossary ensured that there would be uniformity in the resulting data. This task required three days.

When this information was collected and assimilated by the VE department financial analyst, the results were added to the functions on the Argus Chart. A report was then prepared and recommendations for action were made to management. The working Argus Chart was then simplified as shown in Figure 14.14.

As can be seen from the Argus Chart, 43% of the available funds went to "confirm design" and only 14% went to "create design." In the evaluation of results it was decided that this was a poor distribution of funds. With their overall understanding of the engineering process, participants saw this distribution as a source of a number of operational problems. In addition, it was also seen that across-the-board budget reductions would be likely to aggravate the situation and cause additional difficulties in design areas.

It was recognized that "confirm design" was a required function. However, the way the function was performed appeared to offer opportunities for major improvements in productivity and, at the same time, could improve the overall engineering operation and obtain more efficient

Function/Cost Worksheet

Dept. No. _0001_ Dept. Name _MECHANICAL COMP_
Date _10/31/74_ Manager _S. WAREZ_

Item No.	No. Req.	Activity		1 TRANS INFO	2 CREATE DESIGN	3 AUTH. PROG.	4 PREP. PLAN	5 CONF DESIGN	6 EVAL. INFO	7 COLLECT DATA	8 MAKE MODEL	9 MODIFY DESIGN	10 IDENT DOCU	11 MODIFY CRITERIA	12 PROCURE MAT'PL	13 MAINT FACIL	14 ESTAB. CRIT.
1		MANAGER	1300	100	60	40		50	150	40				40			
2		SECRETARY	1736	1000					60	60			60				
3		DESIGN SUPERVISOR	1438	40	40			40	60	40	40		80				
4		DES. ENG SUPERVISOR	1344	20	100	40		60	80	80	40		80	60	60		40
5		DEVELOPM'T SUPER.	2270					80	160	280			280	160	40	40	20	40
6		TECHNICAL SPECIALIST	1078		200	40							60	40			
7		SR. DESIGN ENGR	2790	40	200	40		160	200	120			240	120	120			160
8		JR DEVELIP. ENGR	5404					320	1280	560	120		760	320	80	80	40	160
9		DESIGN ENGR	11109	280	3360	140	280	420	580	560			580	560	560			140
10		DEVELOP ENGR	22560					1200	4200	3000	750		2700	1200	300	1200	1500	600
11		DESIGN LEADER	4908	180	480			180	160				492		180			120
12		MAJOR L.O. & DES. DRAFTSMAN	22204		2320			1040	780				1560	520	240			520
13		MINOR L.O. & DETAILER DFTSMN	17580		4800			800	400				1200	400	100		
14		CLERK	3392	880									880	80			40
15		MECHANIC	26528							4800	8600	6400	640		640	4800
16		MATERIAL	$193000										$19,000		$137000	711,000

SEGMENT OF A TYPICAL DEPARTMENT WEEK: SLT

TOTAL MTL DOLLARS -193,200
TOTAL HOURS 142220 WH 79 PEOPLE

Cost Totals

Figure 14.13 Cost–Function–Value Analysis

use of engineering funds. The immediate recommendation was to review all areas involved in "confirm design" to determine alternate ways to satisfy the requirements at lower cost.

The result was to conduct value studies of major areas involved in "confirm design." In due course, major changes were made in several areas which substantially increased output, and resulted in product cost improvement and the avoidance of major capital investments. Several of the areas affected were engine certification, development dynamometers, and the full-scale wind tunnel.

The same methods and procedures outlined here have been widely used to improve operations in many ways for a number of major companies. For example, a title insurance company is now conducting substantially more business than ever before with half the personnel. A project at the largest machining operation in the United States was able to reduce the amount of natural gas used for plant heating and material heat treating by one third, for savings in the millions of dollars. The result of the analysis of a large mechanical laboratory resulted in a 22% improvement in productivity by laboratory mechanics. A plant manufacturing computer chips was able to reduce the time required to construct new plants by 50%.

15	16	17	18	19	20	21	22	23	24	25	26	27	28	29
CLARITY	SUFFC	CONF.	EVAL.	ORGNS	DEV.	STORE	ADV	EVAL	DET.	CREATE	ESTAB	ALLOC	DIRECT	MANT.
DES.	DWG	INFO	INFO	INFO	INFO	INFO	TECH.	SUGG	DIRECT	ALT	PRIO	RES.	PEOPLE	STAFF
		40	100					50	100	40	100		100	40
			60	156		120								100
40		60	60	40	80	40		20	90	40	200		180	60
		60	40		70		40	90	60	40	90	20	100	20
			100				40	40	80	160	120	40	90	80
			60					400	60		80	40		
40		160	160		80		60	80	120	80	200		210	80
20			160				60	80	160	240	160		320	80
140		280	420	100	280	140	280	420	360	259	140		100	100
300			1200			300	600	300	300	1200		225	600	
120	120	360	240	560	450	120	120	120		240	120		480	
520	520	1560	520	520	3120		520	520	260	492				
980	400	600	400	200	6600		100	100	100	100				
	160			240	240	400								
640														

Figure 14.13 (continued)

Congestion in High School Corridors During Class Change Periods

One of the local value society members came to the chapter and pointed out the problems their school district was having. He asked if we would talk to the school board regarding VE and the use of the process to aid in resolving some of their problems. We did meet with the school board, and they agreed to try our methods. We told them what we would need to conduct a workshop and laid out the ground rules for the program, including the selection of projects and task force team members.

The board selected two projects that they considered to be the most important at the time. The two projects were communication between the school and home, and congestion in the high school corridors during class change periods. At the time of the workshop the school was the second largest in the state.

The projects were to be conducted from 7 to 10 PM one night a week over a six-week period. We wanted to have six people on each team and

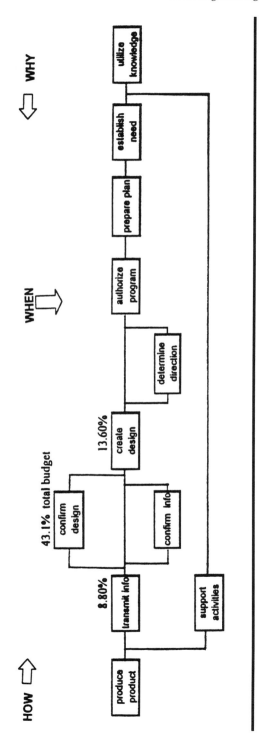

Figure 14.14   Argus Chart — Engineering Department Operations

decided that we should start with two teams on each project to account for dropouts and to ensure an adequate staff during the course of the program. This required at least 24 people to start the program. Team members included teachers, counselors, accountants, psychologists, dentists, policemen, machinists, engineers, attorneys, a principal, PTA members, some of whom were housewives, and a male and female student. In addition, each team was assigned a VE professional to guide them through the process.

The program started with a one-hour orientation covering the process and a discussion of what was expected from each participant. As expected, there were dropouts, but we consolidated the groups without difficulty and were able to complete the program on time with a very enthusiastic group. Recommendations were made to the school board, and many of them were accepted and implemented.

The recommendations made by the teams represented an objective, penetrating analysis of all facets of the problems. Both the school professionals and the community participants received an excellent education in the problems and developed a much greater understanding of each other's wants and needs, for an overall cooperative environment.

In the congestion problem, the example cited here, the team broke the project down into functions; a partial list is shown in Figure 14.15. In the early stage the problem was very hazy; however, as time progressed the problem became clearer until it was possible to zero in on aspects of the problem that caused concern and appeared to cause the greatest confusion. As a result of the function analysis, "allow movement" was believed to create the major part of the problem, and book flow and traffic patterns were deemed to cause the greatest difficulties. These were studied in detail. The student participants proved to be invaluable in that they were able to form the link between the student body and the task force. They conducted several surveys that were useful in deliberations. Figure 14.16 is an abridged Argus Chart of the problem, and Figure 14.17 is a floor plan of the school.

Recommendations were both short- and long-range and included the following:

Book Flow — Book flow causes excessive locker usage. Surveys indicated a constant flow of traffic to lockers to pick up and drop off books. It was recommended to eliminate the need to bring books to class by instead providing a set of classroom texts. The extra book cost would be minimal and in the long run probably reduce costs because assigned books would last longer with less wear and tear.

verb	noun	verb	noun	verb	noun
educate	students	transmit	knowledge	create	climate
group	students	allow	movement	channel	flow
signal	time	establish	identity	encourage	maturation
allow	freedom	permit	alternative	prescribe	program
schedule	classes	establish	requirements	meet	schedule
guide	student	prescribe	behavior	develop	values
test	values	develop	relationships	communicate	ideas
create	confusion	observe	behavior	facilitate	passage
delay	arrival	relieve	tension	seek	haven

Figure 14.15   Function List

Traffic Patterns — Traffic patterns create serious conflict. Traffic studies showed that up to 25 crossover patterns were in the ramp area, the area that connected the public areas to the classroom area. The high-density pattern was caused by the addition of a stairway at the south end of the ramp after the building was erected. The resulting jostling and conflict produced high tension in students and considerable lateness to class. It was recommended that the congestion could be relieved by either enclosing the lower ramp or adding a stairway bypass or by doing both.

Enclosing the lower ramp would improve traffic flow, not only across the ramp but more important, by bypassing the stairway. This would eliminate crossover patterns and reduce congestion. Further additional improvement in the area, such as a lounge and snack bar, could reduce overall traffic in the traffic pattern. Improving this area could change the traffic pattern of the entire building. The possibility of closing the area was discussed with the fire department, and it was believed that a workable arrangement could be developed.

Adding a stairway bypass ramp or a 3- to 4-step stairway constructed through the library coordinator's office would bypass traffic around a critical point. The studies and charts show that the 25 crossover paths could be reduced to about 16. This would mean a one third reduction in crossover patterns, resulting in less conflict, but it would require a capital expenditure which was not available at the time.

Large numbers of students create tension. To reduce this tension and improve overall conditions, additional long- and short-term recommendations were developed in the area of organization and operation of the system. The short-term recommendations included reorganizing to a cluster concept and developing additional free space in the facility. The long-term recommendations involved redistributing the students and a 12-month school year. Short-term recommendations follow:

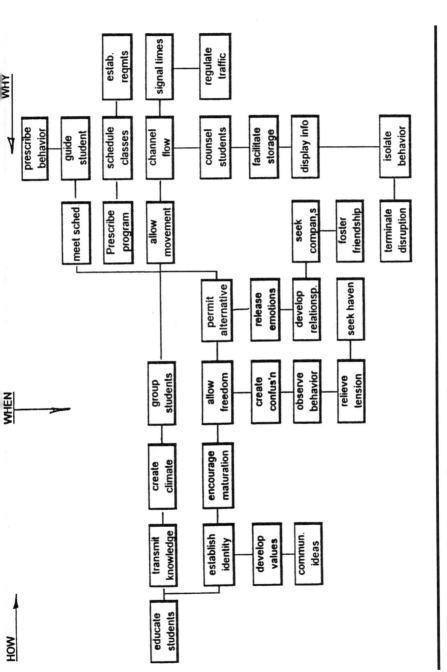

Figure 14.16 High School Congestion — Argus Chart

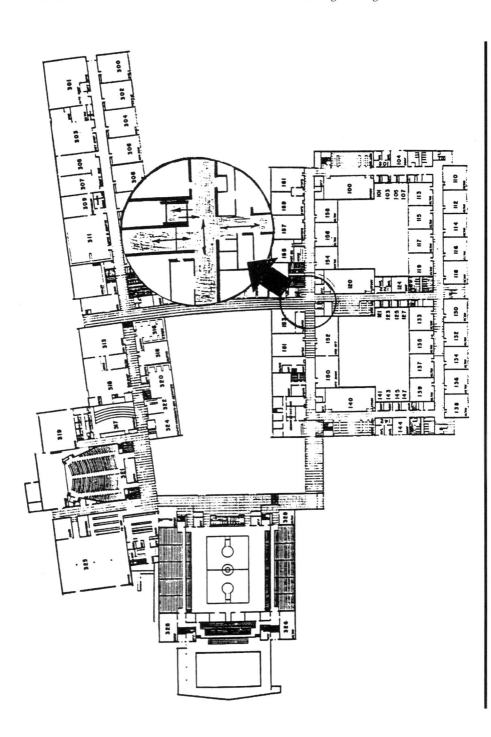

Cluster Concept — Reorganize the school into four semi-self-contained schools (clusters). Each cluster would be complete with its own administrator, teachers, students (800), and facilities (limited). The cluster would include all four grades, but each would be assigned a section of the school for its operations. Students could be redistributed by interest groups or some other acceptable method. Students would mix at shop, gym, or other courses requiring the use of general facilities.

Develop Free Space — Coupled with the clusters would be courtyards or other recreational areas for each cluster. These areas would be furnished with suitable street furniture and other facilities.

These recommendations would reduce long-distance traffic and a major proportion of the hallway traffic. They would also provide a smaller school unit and could be expected to develop a greater sense of identity with teachers, counselors, organizations, etc.

Long-term recommendations:

Redistribute Students — The idea is to reduce the number of people in the building to reduce congestion. Among the ideas was to transfer Grade 9 students to the junior high level, and Grade 7 to the elementary level. Grade 8 would remain in the junior high level. This would redistribute students to less heavily crowded facilities.

Twelve-Month School Year — This would reduce the number of people in the building. It has been tried in other school districts and found to have favorable results. However, there are some objections, such as tradition, vacation adjustments, after-school activities, and the possible necessity to air-condition the buildings. There are also advantages, and among the most important are more efficient use of the buildings and the reduction in congestion.

Career Analysis and Planning

In all probability most people get into careers by accident. How does an engineer, insurance salesman, accountant, or a doctor decide on a career? There are cases where engineers, lawyers, doctors, businessmen, etc., have been in the family for years. A tradition has been set; a dynasty has been created because the son or daughter doesn't want to disappoint Dad. In other cases, people may follow their friends because "everyone is doing it."

So it is that in most cases, whether the person is a student, engineer, secretary, mechanic, plumber, clerk, toolmaker, or in any other job, there is confusion as to what makes up the job. People do not really know what they want to do because they do not know what the job entails. In fact, it is not unusual to find that many managers do not know what many of their subordinates do.

The difficult question is, what do you want to do? However, the answer may be easier if the question is reversed. Ask yourself, "What have I done? What are my accomplishments?" This can then become the basis for a rational decision.

Identifying a career path requires identifying accomplishments. After this has been done it is possible to identify your special skills. The combination of accomplishments and the identification of the skills to achieve them is the key to a successful career-change campaign.

An accomplishment is different from a duty. Most jobs list duties. An accomplishment is the successful result of an action or the attainment of an objective. If you built the Brooklyn Bridge, say, "I built the Brooklyn Bridge." This is an accomplishment. One of your accomplishments may have been to organize and coordinate the moving of a plant to a new location. Say, "I developed the plans and directed the move."

It is not an easy task to define your accomplishments. Help is frequently required. It appears that highly skilled people often have difficulty identifying accomplishments. This is probably because they perform many difficult tasks in a routine manner and don't recognize the importance of these tasks to others.

Although the job of identifying accomplishments is difficult, some consolation may be found in the fact that everyone has difficulty. Engineers, architects, bank presidents, housewives, teachers, all have difficulty. However, the job is not impossible once you have made up your mind to start. Get a large pad and a dozen pencils and begin.

The first step is to prepare a complete list of your past accomplishments. List everything you can think of, from your success in selling Girl Scout cookies or Boys' Life magazines to the record-breaking construction project you completed. List all job duties, activities, awards, acknowledgments, any and everything you can think of. List community activities, church, college, and high school projects if possible. Use short statements and don't be too critical. This will form the basis for your self analysis.

The Problem — Several years ago a young man came to me with the problem that he had been working as an advertising account executive for a over a year and had not seen much advancement or opportunity for the future. He asked if I had any suggestions as to how he might improve his career. We applied the VE Job Plan to his problem and achieved an extremely successful result.

The first step, identifying accomplishments, included the following examples:

Made sales presentations	Developed new accounts
Made market surveys	Maintained records
Supervised program progress	Negotiated agreements

A person should be able to list hundreds of separate items. The older and more experienced person will have a longer list, but the recent graduate will also have a substantial list. The list must now be screened to eliminate duplication and clarify the items. A convenient way to do this is to identify the verb–noun apparent function in each item. This forms the basis for a function list. Defining the functions is the key to a successful project. These two examples are from the account executive's list of accomplishments and shows how the functions were developed from the list of accomplishments.

Accomplishments — Advertising Account Executive
Obtained commitment from client by *making presentations* based on performance and *buyer's needs* to *attract and increase sales.* *Located clients* by *evaluating the market* to *determine potential* for sales effort.

The functions are identified by selecting the two-word verb–noun combinations. The selected combinations are listed below. Notice that in some cases the combinations have been revised.

Verb	Noun	Verb	Noun
obtain	commitment	evaluate	market
make	presentations	increase	sales
satisfy	needs	locate	clients
attract	attention	determine	potential

In total, 28 functions were defined from the final list of accomplishments, and an Argus Chart was developed. A segment of the Argus Chart is shown in Figure 14.18. The completed chart identifies accomplishments, and tells why and how they were performed. The chart was then evaluated

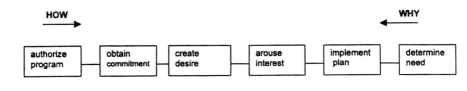

HOW →   ← **WHY**

| authorize program | obtain commitment | create desire | arouse interest | implement plan | determine need |

Figure 14.18   Advertising Account Executive

by applying a combination of two simple methods. The first was to identify likes and dislikes and best results. The second method was Pareto Voting and Paired Comparisons. Some attractive features were listed as:

personal contact	likes associates
create programs	good location
develop programs	good experience
implement programs	experience results
job freedom	

The main dislike was project research.

Pareto Voting and Paired Comparisons were applied to the Argus Chart, and results were ranked and weighted. The five functions that ranked highest are shown below.

Rank	Function	Weight
1	achieve success	30-C
2	obtain commission	22-M
3	negotiate agreement	16-N
4	get commission	12-B
5	satisfy customer	11-A

A job description was developed based on the function list. For example:

What	How	Why
Get commission	Create desire	Authorize program
	Obtain interest	Implement program
	Arouse interest	Complete project
	Implement plan	Present results
	Determine need	Meet needs
		Satisfy customer

From this information, a specific accomplishment was prepared. For example,

> I *obtained commitments* from clients for advertising services/by making sales *presentations* based on an *understanding* of their needs and a record of past company performance to *authorize* a new *program*.

Now, what was the specific benefit to the company, to the client, and to yourself?

Added 12 new accounts
Made a major contribution to company income
Increased confidence and developed credibility in the marketplace

After this type of analysis was made for as many functions as necessary, all of the information required to make an effective sales presentation was available.

There are as many different types of sales presentations, or in this case, resumés, as there are people. There is probably no one form that will satisfy everyone, so don't worry about it. However, you must keep in mind the five key areas of management: planning, implementation, supervision, directing, and coordinating. The example shown in Figure 14.19 is the result of the analysis. Notice the simple straightforward presentation. One clear-cut, interesting accomplishment after another attracts attention. Next comes supporting details, and the third paragraph highlights background details. There were a number of compliments made on the presentation, which resulted in several offers of employment. The resumé obtained the interviews, and the knowledge gained in constructing the Argus Chart greatly aided in the interview.

The rest of the presentation includes a list of past employers, education, outside interests, and vital statistics, all of which are facts.

The result of this project was a very successful campaign that resulted in a new position with one of the largest advertising companies in the country.

Small Parts Analysis

The previous examples have all been of relatively large and complex projects. However, VE and the disciplines of function definition and function analysis can also be invaluable on small, seemingly commonplace problems. This example of small parts analysis involves a cost analysis of

SUMMARY OF EXPERIENCE OF;

GEORGE T. SMITH
1000 OAK STREET
DOWNTOWN, MI 00000
TEL. 000 - 000 - 0000

POSITION DESIRED

Account Supervisor or Advertising Director

PAST RECORD

Account Supervisor: In the past two years as an Account Supervisor with Dynamic Associates, Inc. I have added 12 new active billing accounts to Dynamic's present client list. These accounts are presently contributing a major percentage of company income accounts.

Account Director: Besides selling these accounts, I analyzed each new account's standing in the industry, evaluated his present advertising activities and planned a program for future action. I was directly involved in the production of all work done for each client in addition to studying and placing all media for the account.

Technical Background: My success has been due to my education and previous experience with industrial and retail businesses as well a on the job training. One unique activity that has contributed to my ability to identify, analyze and solve problems has been attendance at the Society of American Value Engineer's Saturday morning FAST Breakfast sessions. These sessions offered the opportunity to obtain experience in the application of Value Engineering techniques to a wide variety of projects from products and services to community affairs.

These experiences have enabled me to work with my company's administrative and creative staffs in addition to my client to provide the professionalism needed to produce effective results.

EXAMPLES OF ACCOMPLISHMENTS:

I joined Dynamic on September 2, 19XX, and as a representative in the Account Supervision Department, I actively began pursuing new accounts. From a basis of no accounts I procured information of many companies which seemed to be potential clients.

My first project as with a local truck tarpaulin company. The problem was to promote their product within a limited budget to expand markets.   I supervised all photography and design for this project besides writing all copy.   The project was then implemented with very successful results culminating in a substantial increase in sales for my client.

Figure 14.19   Advertising Account Executive — Resumé

a computer frame made up of a number of small stampings. A review of the stampings by a value engineer produced substantial cost savings for the overall project by simply reviewing the Tests for Value in the Appendix and applying function definition and creative thinking.

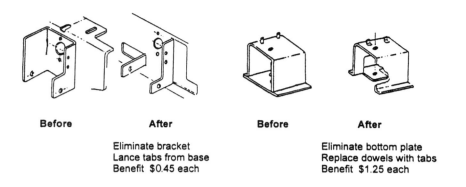

Before	After	Before	After
	Eliminate bracket Lance tabs from base Benefit $0.45 each		Eliminate bottom plate Replace dowels with tabs Benefit $1.25 each

Figure 14.20   Computer Frame Components

For example, for one of the parts in Figure 14.20, the bracket was a separate stamping attached to the frame. The function of the bracket was determined, and creative thinking resulted in the tabs being lanced out of the frame for a substantial saving in cost. In the second example in Figure 14.20, the dowels were an expensive part of the assembly. Again the function was determined, and the result was lancing tabs out of the component to satisfy both the dowel requirement and base connection. The function was satisfied in both cases at a saving in material and labor. The answer is to Think Function and apply creative thinking to develop new ways to achieve the function.

APPENDIX

Appendix

Table of contents

The Job Plan

1. Information Phase

What is it?

Collect all data, drawings, blueprints, costs, parts, flow sheets, process sheets, or other pertinent information. Talk with people. Ask questions, listen, develop. Discuss all details, discuss and probe to become thoroughly familiar with the project. Set goal for achievement.

What does it do?

Define functions.
    Determine basic function(s).
Construct Argus Chart (FAST Diagram).

What does it cost?

Function–cost analysis.

What is it worth?

Establish a value for each function.
Determine overall value for the product or service.
Compare overall value to goal for achievement.

Where is the problem?

Analyze the Argus Chart.
    Locate poor value functions.

What can we do?

Pinpoint the areas for creativity.
    Targets for opportunity (TFO).

2. Creative Phase
What else will do the job?

Create alternative ways to perform the poor value functions. Apply creative techniques. Develop unique solutions. Combine or eliminate functions. Brainstorm, look for revolutionary ideas. Don't overlook possible serendipitous discoveries. **Do not apply any constraints to thinking at this stage.**

3. Evaluation Phase
Select the best ideas

Screen **all** creative ideas.
 Evaluate carefully, looking for useful solutions.
 Combine the best ideas.
 Categorize into basic groups. Screen for the best ideas.

How much does it cost?

Generate relative costs.
Analyze potential.
Anticipate roadblocks.

4. Planning Phase
Develop best ideas

Develop practical solutions.
 Obtain accurate costs.
 Review engineering and manufacturing requirements.
 Check quality, reliability. Talk with people.
Resolve anticipated roadblocks.
Develop an alternative solution.
Plan your program to sell.
Show the benefits.

5. Reporting Phase

Present ideas to management

Show before and after costs, advantages and disadvantages, nonrecurring costs of development and implementation, scrap, warranty, and other forecasts and net benefit.

Plan the program to sell.

Don't tell them anything they don't want to hear.

6. Implementation Phase

Be certain the change satisfies the original intention.
   Audit actual costs.

Key Techniques

1. Define functions
2. Identify and overcome roadblocks
3. Use specialty products and services
4. Bring new information
5. Construct Argus Chart
6. Cost and evaluate the Argus Chart
7. User accurate costs
8. Establish goals
9. Get all information from the best source
10. Use good human relations
11. Get all the facts
12. Blast–Create–Refine
13. Put a dollar sign on key tolerances
14. Put a dollar sign on the main idea
15. Use your own judgment
16. Spend the company's money as you would your own
17. Use the company's services
18. Work on specifics not generalities
19. Use standards
20. Use imagination
21. Challenge requirements
22. Use supplier services

Techniques vs. Job Plan

1. Information Phase   1, **2**, 4, 5, 6, 7, 8, 9, **10**, 11, 13, 14, **15, 16**, 17, **18**, 19, 21, 22,

    What is it?
    What does it do?
    What does it cost?
    What is it worth?
    Where is the problem?
    What can we do?

2. Creative Phase   10, 12, **15**, 20, 21
    What else will do the job?
    Use creative techniques
    Brainstorm
    Create alternatives

3. Evaluation Phase   **2**, 3, 7, 8, **10**, 11, 13, 14, **15, 16**, 17, **18**, 19, 20, 21, 22

    Review suggestions
    Refine results
    Evaluate carefully

4. Planning Phase   **2**, 3, 5, 7, 8, **10**, 11, 13, 14, **15, 16**, 17, **18**, 19, 20, 21, 22

    Develop best ideas
    Develop alternative solution
    Plan program to sell

5. Reporting Phase   **2**, 10, **15, 16**, **18**
    Present ideas to management.
    Make recommendation for action
    Ask for approval

6. Implementation Phase   2, 10, 15, 16, 17, **18**, 19, 20, 22
    Monitor to assure proper
      implementation

Bold numbers indicate techniques that apply at every phase of the Job Plan as well as in most other activities.

Argus Chart Construction

1. **List all functions**
2. **Determine basic function**
3. **Construct primary path**
4. **Determine when remaining functions happen to start dependent family tree(s)**
5. **Develop supporting primary path for family tree(s)**
6. **Complete Argus Plot and check how-why relationships**
7. **Complete Plot**
   Number functions
   Develop function cost
   Develop function value
   Determine poor value functions. Targets for Opportunity (TFO)
   Determine goal for achievement and compare to original target
8. **Apply creative techniques**
   Develop alternative solutions

Tests for Value

1. Can we do without it?
2. Does it need all of its features?
3. Does it cost more than it's worth?
4. Is there something better that can do the job?
5. Can it be made by a less costly method?
6. Can a standard item be used?
7. Does it cost more than the total of reasonable costs for material, labor, burden, and profit?
8. Can a less costly tooling method be used, considering the quantities involved?
9. Can another dependable supplier provide it for less?
10. Is anyone buying it for less?
11. Would you pay the price if you were spending your own money?

Screening and Selection Techniques

1. Paired Comparisons/Numerical Evaluation Technique

Used when it is desired to determine the best alternative or the relative weighing and/or ranking of a number of alternatives. Each alternative is in turn compared to every other alternative and a rating established depending on the length of time required to make the decision of relative importance. The technique is one of the simplest of the ranking/weighting evaluation systems and is very effective.

A. E. Mudge, Numerical evaluation of functional relationships, *Proceedings,* Society of American Value Engineers, 1967.

2. Direct Magnitude Estimation

Used when it is desired to determine the best alternative or the relative weighing and/or ranking of a large number of alternatives. Each group member is asked to rate the relative importance of alternatives based on any comparative magnitude system desired. The system is ideal for computerization. Comparisons are based on geometric means and normalized for direct comparison on a common scale.

D. M. Meyer, Direct magnitude estimation, *Proceedings,* Society of American Value Engineers, 1971.

3. Pareto Voting

Used to reduce a large number of alternatives to a more manageable number and then achieve ranking by the use of other techniques. Based on Pareto's Laws of Maldistribution or the 80–20 concept.

M. Larry Shillito, Pareto voting, *Proceedings,* Society of American Value Engineers, 1973

4. Ranking and Weighting

Used to compare alternatives on a common scale of measurement. Ranking may be accomplished by simply listing in sequence of preference. Weighting is accomplished by use of an index number of relative importance when compared to objective goals. More sophisticated techniques are described in the article.

J. Christopher Jones, *Design Methods, Seeds of Human Futures,* John Wiley & Sons, New York, 1970, 377.

5. Criterion Function Analysis

Used to aid in selection and optimization of a group of variables. The system makes use of mathematical formulas which utilize all of the important criteria of the problem.

R. S. Schermerhorn and M. I. Taft, Measuring design intangibles, *Machine Design*. December 19, 1968.

6. Weighted Constraints

Used for preliminary or rapid approximate ranking or screening of alternatives. For screening, alternatives are classified into categories; comparison is made by feature comparisons and/or establishing advantages and disadvantages; and if desired, ranking can be accomplished by the use of a direct and simple comparative rating system such as 1–10.

Department of Defense, *Principles and Applications of Value Engineering,* Volume 1, U.S. Government Printing Office, Washington, D.C., 1968.

7. Cost Function Analysis

Used to determine the extent to which alternatives can perform a function and the cost involved. Relates cost to function in a way that makes evaluation of the function value possible and permit comparisons to be made. Part of the Argus System. (Function Analysis System Technique, FAST)

Chrysler Corporation, *Cost, Function, Value Analysis, Value Control Manual,* R. J. Park and Associates Inc., 1978.

8. Criteria Analysis

Used to evaluate the performance of design alternatives in new product design. It can be used in the analysis of consumer products characterized by aesthetic as well as functional value. The system makes use of function matrices to graphically show the relationship of components to function.

D. J. DeMarle, Criteria analysis of consumer products, *Proceedings,* Society of American Value Engineers, 1971.

9. Cluster Analysis

Used to determine the best combination of choices among a number of alternatives rather than the one best choice. The intent is to determine the balanced combination that will satisfy a larger number of market requirements.

M. L. Shillito, Cluster analysis, amplification of the value index, *Proceedings,* Society of American Value Engineers, 1974.

10. Combinix

Used in decision making by comparing alternatives with the use of the Combinix Scoreboard. The Scoreboard provides a matrix for combining the benefits that make up a requirement and for noting their relative importance so that various courses of action or choices may be selected.

C. Fallon, *Value Analysis to Improve Productivity,* Wiley Interscience, 1971. 2nd edition, Triangle Press, Irving, TX.

11. Decision Analysis

Used to quickly reduce alternatives by classifying according to Musts and Wants. Those not meeting Musts are discarded. Ranking and/or weighting criteria may then be applied to surviving alternatives.

C. H. Kepner and B. B. Trego, *The Rational Manager,* McGraw-Hill, New York, 1965.

12. Delphi Method

Used to evaluate subjective alternatives by asking a group of experts or specialists to predict results. The objective is to combine the differing opinions of the participants into a single position by systematic arbitration.

R. S. Schermerhorn and M. I. Taft, Measuring design intangibles, *Machine Design,* December, 19, 1968, 108.

13. Checklists

Used to bring previous knowledge to bear to ensure that important considerations are properly evaluated and not overlooked in making a decision.

J. Christopher Jones, *Design Methods, Seeds of Human Futures,* John Wiley & Sons, New York, 1970, 367.

Idea Needlers or Thought Stimulators

How much of this is the result of custom, tradition, or options?

Why does it have to be this shape?

How would I design it if I had to build it in my own workshop?

What if it were turned inside out, reversed or turned upside down?

What if it were larger, higher, wider, thicker, lower or longer?

What else can it be made to do?

Suppose this was left out?

How can it be done piecemeal?

How can it appeal to the senses?

How about extra value?

Can this be multiplied?

What if this was blown up?

What if this was carried to extremes?

How can this be made more compact?

Would this be better symmetrical or asymmetrical?

In what form could this be, liquid, powder, paste, or solid?

Could it be a rod, tube, triangle, cube, or sphere?

Can motion be added to it?

Will it be better standing still?

What other layout might be better?

Can cause and effect be reversed?

Is one possibility the other?

Should it be put on the other end or in the middle?

Should it slide instead of rotating?

Demonstrate or describe by telling what it isn't.

Has a search been made of the patent literature or trade journals?

Could a vender supply this for quicker assembly?

What other materials will do the job?

What is similar to this but costs less? Why?

What if it were made lighter or faster?

What motion or power is wasted?

Could the packaging be used for something afterward?

If all specifications could be forgotten how could the basic function be accomplished?

Could these be made to meet specifications?

How do competitors solve problems like this?

Creative Games

1. Using four straight lines, divide the area of a circle into the maximum number of parts.

2. Arrange 12 trees in such a way as to have six straight rows of four trees each (two solutions).

3. With four straight lines, go through each of the nine dots without removing the pencil from the paper.

4. How many squares do you see?

5. Can you connect like numbers without crossing lines or leaving the perimeter of the figure?

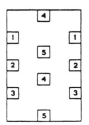

6. Copy this design and cut it into three parts with two straight cuts so that the individual parts, when reassembled, will form a square.

Coin Shift

This classic parlor puzzle illustrates the rigidity with which most people approach problem situations. While, traditionally, only one solution is thought to be possible for this problem, you can come up with several if you use your imagination.

Set a dime between two quarters, with its edges touching both. Now get the right hand quarter into the middle position without moving the dime or touching the left hand quarter.

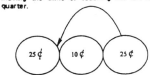

Answers

1. The answer is 11. Those using symmetry of wheel spokes will not get it. This problem examines several characteristics — sensitivity (realizing that the obvious won't do), creative discontent, and flexibility.

2. This problem examines imagination — various configurations being visualized, and flexibility — getting a second solution without being "hung up" on the first.

3. This problem cannot be solved by remaining within the confines of the nine dots. The characteristic of creative discontent is examined by this problem.

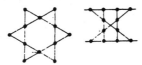

Coin Shift

1. With your left index finger, press firmly on the dime. With the right index finger (or two fingers), slide the right-hand quarter to the right, then strike firmly against the dime. The left-hand quarter will spring aside. Move the right-hand quarter into the exposed space.

2. Press your fingers tightly on the dime and the right-hand quarter. Blow at the left-hand quarter and it will move aside.

4. There are 30 squares (sixteen 1 x 1; nine 2 x 2; four 3 x 3; and one 4 x 4). This is a test of sensitivity, realizing that more than just the 16 small squares exist.

5. The tendency is to work the problem by connecting the squares in numerical sequence. A person with constructive discontent will not be restricted by the relationship of the number.

6.

3. Hold the dime and the right-hand quarter firmly in place. With your left knee, lift the right-hand side of the table sufficiently for the left-hand quarter to slide away.

4. Place the left-hand quarter on a piece of paper so half of it extends beyond the paper. Then move the paper with the quarter to the left.

Sample Report to Management

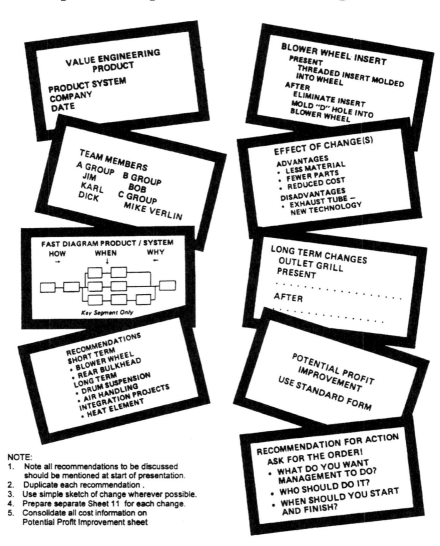

VALUE ENGINEERING
PRODUCT

PRODUCT SYSTEM
COMPANY
DATE

TEAM MEMBERS
A GROUP   B GROUP
JIM
KARL   BOB
DICK   C GROUP
          MIKE VERLIN

FAST DIAGRAM PRODUCT / SYSTEM
HOW   WHEN   WHY
→   ↓   ←

Key Segment Only

RECOMMENDATIONS
SHORT TERM
• BLOWER WHEEL
• REAR BULKHEAD
LONG TERM
• DRUM SUSPENSION
• AIR HANDLING
INTEGRATION PROJECTS
• HEAT ELEMENT

BLOWER WHEEL INSERT
PRESENT
    THREADED INSERT MOLDED
    INTO WHEEL
AFTER
    ELIMINATE INSERT
    MOLD "D" HOLE INTO
    BLOWER WHEEL

EFFECT OF CHANGE(S)

ADVANTAGES
• LESS MATERIAL
• FEWER PARTS
• REDUCED COST

DISADVANTAGES
• EXHAUST TUBE —
  NEW TECHNOLOGY

LONG TERM CHANGES
OUTLET GRILL
PRESENT
. . . .

AFTER
. . . .

POTENTIAL PROFIT
IMPROVEMENT
USE STANDARD FORM

RECOMMENDATION FOR ACTION
ASK FOR THE ORDER!
• WHAT DO YOU WANT
  MANAGEMENT TO DO?
• WHO SHOULD DO IT?
• WHEN SHOULD YOU START
  AND FINISH?

NOTE:
1. Note all recommendations to be discussed
   should be mentioned at start of presentation.
2. Duplicate each recommendation .
3. Use simple sketch of change wherever possible.
4. Prepare separate Sheet 11 for each change.
5. Consolidate all cost information on
   Potential Profit Improvement sheet

Workshop Schedule

Time	Day 1	Day 2	Day 3	Day 4	Day 5
8:00 AM	Welcome Introduction to VE 8:15 – 9:00	Function Evaluation 8:00 - 8:30 Function – Cost – Analysis 8:30 – 9:00	Purchasing Techniques 8:00 – 9:00	Workshop 8:00 – 11:00	Final Report Preparation 8:00 – 11:30
9:00	Coffee Break Listening Is Good Business 9:15 – 10:00	Coffee Break Workshop 9:15 – 11:00	Coffee Break Workshop 9:15 – 11:00	Coffee Break Workshop continued	Coffee Break Workshop continued
10:00	Definitions and Job Plan 10:00 – 10:45				
11:00	Information Phase Project assignment 10:45 – 11:30	Argus Chart Review 11:00 – 11:30	Unnecessary Cost 11:00 – 11:30		
12:00 M	Lunch 11:30 - 12:15 Goal For Achievement 12:15 – 12:45	Lunch 11:30 - 12:15 Creative Phase Creativity 12:15 – 1:45	Lunch 11:30 - 12:15 Human Relations Group Dynamics 12:15 – 1:15	Lunch 11:30 - 12:15 Reporting Phase 12:15 - 1:00	Lunch 11:30 -12:15 Workshop 12:15 – 1:30
1:00	Human Relations Habits and Attitudes 12:45 – 1:30 Cost Visibility 1:30 – 2:00		Workshop 1:15 – 2:30	Workshop 1:00 – 4:30 Roadblocks - Tape	Report to Management 1:30 – 2:30
2:00	Function Analysis 2:00 – 2:30	Workshop 1:45 – 2:30			
3:00	Coffee Break Workshop 2:45 – 3:15	Coffee Break Blast – Create – Refine 2:45 – 3:00	Coffee Break Workshop 2:45 – 4:45	Coffee Break Workshop	Coffee Break Questions and Answers 2:30 – 3:30
4:00	Argus Charts 3:15 - 3:45 Workshop 3:45 - 5:00	Workshop 3:00 – 4:45			Critique and Presentation of Certificates 3:30 -4:00
5:00		Evaluation Phase	Planning Phase	Implementation Phase	

PRELIMINARY   October 1, 1997

Cost and Inflation — The Effect on Our Livelihood

If you had been born around 1950 costs and prices were quite different than they are today. An engineer started on his first job for about $2,400.00 annually and many people worked in factories and on clerical jobs for from fifteen to twenty cents an hour. However, when looking back we must consider the inflation rate to determine what the equivalent numbers would be at today's rate. In the January 1997 Kiplinger's Personal Finance magazine an article entitled "Then and Now" lists a number of items with their original cost and its 1997 equivalent and the actual 1997 cost. Several items were taken from the article and are shown here for your comparison. The percent of the present cost compared to the original cost in today's dollars is also shown.

Although the list is based on averages and medians it shows that some items are up and others down. We have reduced the price of food appreciably but clothing has remained about the same. Transportation varies; the subway is up but airlines are down and automobiles are about the same. However, the appliance industry has reduced the cost of their products appreciably. The data indicate that it is important to know what your priorities are because they can strongly affect your budget.

Incomes have about doubled as has the income tax; however, special taxes such as Social Security have increased appreciably and many new taxes such as sales and excise taxes have been added.

We may complain about the high cost of things but we must also consider content. In 1950 a twelve inch B&W television set cost $445.00 which in equivalent 1997 dollars was $1,470.00. Today a 25 inch color stereo TV costs just $300.00, 9% of the 1950 cost for the smaller B&W unit.

FOOD

Gallon of milk, loaf of bread, dozen eggs,
pound of butter, pound of ground beef
1947: $3.10   ($23.00)
1997: $8.00   (35%)

Cup of coffee
1947:   5 cents   (37 cents)
1997: $1.27   (343%)

McDonald's hamburger
1955: $.15   ($.90)
1997 $.59   (65%)

CLOTHING

Brooks Brothers suit
1947: $70.00   ($520)
1997: $598.00   ($115%)

SHELTER

Median home price
1950: $7,350.00 $49,330.00)
1996: $131,500.00   (267%)

Automatic Clothes Washer
1947: $240.00   ($1,720)
1997: $270.00   (16%)

Refrigerator
1947: $200.00   ($1,470) 9 cu. ft.
1997: $700.00   (48%) 20 cu. ft.

Television
1947 $445.00   ($3,280.00) 12 inch B&W
1997: $300.00   (9%) 25 inch color-stereo

TRANSPORTATION

NYC Subway
1947:   10 cents   ($.75)
1997: $1.50   (200%)

Top Selling Automobile
1947: $1,220.00   ($20,410.00) Chevrolet
1997: $18,000.00   (90%) Ford Taurus

Airline Fare — Kansas City to Paris
1947: $674.00   ($4,770) 37 hours
1997: $740.00   (16%) 10.5 hours

ENTERTAINMENT

Disney Land — Adult admission
1956: $1.00   ($7.00)
1997: $34.00   (485%)

NY Yankees — Lower box seat
1947: $2.50   ($18.00)
1997: $19.00   (105%)

Movie
1947: $.70   ($4.45)   Two admissions
1997: $13.00 (290%)   Two admissions

INCOME AND TAXES
Median Family Income
1947: $3,000.00   ($19,700.00)
1997: $41,000.00   (208%)
Average auto worker Income
1947: $2,770.00   ($20,410.00)
1997: $39,000.00   (191%)
Average Teacher Income
1947: $2,640.00   ($19,450.00)
1997: $37,850.00   (195%)

EDUCATION

NYC Schools — pupil expenditure
1947: $263.00   ($1,940.00)
1997: $8,300.00   (428%)
Yearly Tuition — Harvard
1947: $525.00   ($3,870)
1997: $19,770.00   (510%)

Yearly Tuition – University of Iowa
1947 $130.00   ($960.00)
1997 $2,470.00   (257%)

MISCELLANEOUS
Winning the Masters
1947: $2,500.00   ($18,420.00)
1997: $450,000.00   (2,442%)

Basic Data excerpted from Then and Now, Melynda Dorel Wilcox, *Kiplinger's Personal Finance* magazine, January 1997, pages 124-126. Percentage increase in cost to 1998 prices developed by R. J. Park

Park's Catalog Of Frequently Used Functions In Value Engineering*

R.J. Park and Associates Inc.
Resource Management through Value Engineering
Birmingham, Michigan

Foreword

The intent of this monograph is to provide a catalog of functions that have been used successfully in a wide range of projects. It is not intended that these functions be copied but rather that the list be used as a memory jogger.

It is also expected that constant reference to the list will help to define the short, terse process and will help to develop the knack of defining functions on the job during daily activities.

Function definition is a process that helps to break down constraints to understanding. It is not simply an exercise. It is a process that requires the most difficult type of work – thinking. If a value study is to be successful it is necessary that this effort be expended.

The monograph is organized in three parts.

1. Brief discussion of function definition.
2. Sample functions listed by ten general categories.
3. A verb–noun list.

The total number of verbs in the list is 277 and the total number of nouns is 446. This is a list of functions used on major projects over a 20 year period. From this it seems possible to draw the conclusion that this simple list provides the material to define most problems we can expect to encounter.

---

* Adapted and reprinted from R. J. Park, *Park's Catalog Of Frequently Used Functions In Value Engineering*, R. J. Park and Associates, Inc., Birmingham, MI, 1982. With permission.

Contents

1. FUNCTION DEFINITION – THE OBJECTIVE

OBJECTIVE — Function definition is a forcing technique that requires consensus among members, eliminates confusion, creates in-depth understanding of the requirement, clarifies overall knowledge of the project and ultimately breaks down the barriers to visualization so necessary to help define the creative questions that will lead to new, outstanding solutions to the project.

In his book, *Techniques of Value Analysis,* Mr. Miles recognized the difficulty of applying this technically simple concept. He said, "While the naming of functions may appear simple, the exact opposite is the rule. In fact, naming them articulately is so difficult and requires such precision in thinking that real care must be taken to prevent abandonment of the task before it is accomplished". He also said, "Intense concentration, even what appears to be overconcentration of mental work on these functions, forms the basis for unexpected steps of advancement of value in the product or service".

From the beginning, it was recognized that defining functions is a difficult task and requires considerable effort to successfully define the functions of a product or service. However, if the goal of maximum creative opportunity is to be achieved, the effort must be expended.

Function analysis is designed to break down the barriers to visualization and to offer outstanding opportunities for creativity. However, to make the system work effectively it is necessary to understand the term *function* as it is intended in Value Engineering.

DEFINITION — Function is defined as the property that makes something work or sell. It is what we pay for. Function represents our needs, desires or requirements depending on the point of view or the role being played by the analyst. The definition of function requires skill, practice and an awareness that function must be defined in a manner that will broaden the opportunity for creativity.

A function is not an action; it is an objective of an action. For example, "file papers" is an action. What is needed is to understand why the action is being performed. What is the objective? It may be "store information". This then could be a proper definition for the function being performed. It should be obvious that there can be a number of actions that take place in order to achieve the "store information".

2. CATEGORIES

The ten categories are a general breakdown of the project spectrum. Each category offers a list of projects and a sample of the functions used in the project.

It can be seen that the same functions appear over and over again although they may have a different meaning in each case depending on the viewpoint of the team or the role being played at the time. There is no attempt to restrict role playing as the construction of the FAST diagram tends to organize the functions by role. Customer, manufacturer, salesperson, owner, etc.

Although only a sample list has been given for each project a complete list of all verbs and nouns used in major projects is provided in Section 3 Verb-Noun list. This list is complete except that a number of poorly defined functions that do not meet the criteria for function definition have been purged.

In preparing this monograph it was interesting to note how the definitions changed over the years as more skill and practice were developed. It is also possible to see this transformation in teams that have worked together on a number of projects. They begin to become aware of the type of definition that is required to produce creative results.

Although there are hundreds of projects that have been included here it is believed that the defined list is quite comprehensive and will adequately cover most needs.

1.0 Administration/Management Information Systems

1.1 Engineering Operations

Transmit	information	establish	criteria
Create	design	maintain	staff
Clarify	design	maintain	facilities
Confirm	information	reconcile	differences
Identify	deviations	develop	plan
Evaluate	suggestions	establish	priorities

1.2 Budget Control/Cost Analysis

measure	performance	discharge	obligations
highlight	deviations	maintain	history
prepare	plan	administer	policy
allocate	resources	develop	procedures
determine	requirements	guide	management
negotiate	agreement	analyze	data

1.3 Business Plans

increase	awareness	improve	productivity
reduce	cost	increase	return
develop	skills	raise	funds
define	goals	recognize	opportunities
obtain	recommendations	increase	standardization
increase	sales	reduce	warranty

1.4 Insurance Claims Adjustments

transmit	information	receive	documentation
evaluate	information	request	payment
make	decision	interpret	policy
analyze	data	protect	investment
request	information	maintain	history
accept	direction	set	priority

1.5 Inventory Simplification

maintain	inventory	improve	service
verify	quality	identify	management
forecast	requirement	update	records
ensure	supply	develop	standards
maintain	equipment	make	comparison
group	parts	modify	parts

1.6 Long Range Plans

track	material	transport	material
procure	material	define	change
maintain	schedule	disperse	material
evaluate	market	create	image
develop	alternatives	verify	direction
establish	checkpoints	determine	requirements
determine	need	generate	capital
meet	complete	develop	concept

1.7 Material Procurement Information Systems

fabricate	components	define	limits
obtain	material	specify	performance
transmit	information	define	appearance
identify	sources	limit	impurities
classify	designs	define	recipe
set	standards	specify	process

1.8 Parts Identification

service	operation	identify	part
fill	requirement	determine	characteristics
maintain	inventory	create	groups
receive	part	minimize	requirements
store	part	allow	flexibility

1.9 Material Procurement System

obtain	material	evaluate	resources
allocate	resources	develop	personnel
identify	priorities	identify	problems
expand	information	review	schedules
identify	need	negotiate	agreement
authorize	expenditure	define	objective

2.0 Capital Equipment Product Analysis

2.1 Automotive Test Equipment — Radiator Cooling Test

maintain	facility	confirm	design
protect	personnel	maintain	pressure
control	environment	employ	utilities
remove	heat	create	structure
direct	operations	direct	flow

2.2 Industrial Air Compressors

create	path	absorb	vibration
increase	pressure	supply	power
flow	air	conduct	current
remove	heat	guide	fluid
reduce	friction	allow	access

2.3 Heavy Duty Trucks

create	torque	resist	bending
apply	force	ensure	convenience
support	weight	resist	corrosion
protect	operator	identify	product
absorb	impact	control	noise

2.4 Stamping Press

guide	force	maintain	contact
adjust	opening	allow	adjustment
absorb	shock	conduct	fluid
resist	bending	protect	operator
reduce	friction	transmit	information

3.0 Government/Community Affairs

3.1 Church Planning

develop	commitment	attract	volunteers
promote	stewardship	maintain	facility
educate	people	hold	series
service	community	worship	God
welcome	strangers	share	beliefs

3.2 Juvenile Delinquency

protect	children	understand	problems
resolve	conflict	effect	change
involve	community	enhance	life
administer	justice	enhance	growth
interpret	laws	adjust	attitude

3.3 Public Works — Lake Dredging

add	energy	obtain	time
create	mixture	hold	mixture
remove	solids	separate	solids
settle	mixture	remove	water
create	differential	build	lagoon

3.4 Youth Assistance Programs

obtain	conformance	identify	clients
direct	behavior	indicate	alternatives
plan	treatment	engender	hope
diagnose	problems	obtain	conformance
show	concern	socialize	people

3.5 Public Works — Lake Dredging

add	energy	obtain	time
create	mixture	hold	mixture
remove	solids	separate	solids
settle	mixture	remove	water
create	differential	build	lagoon

3.6 Youth Assistance Programs

obtain	conformance	identity	client
direct	behavior	indicate	alternatives
plan	treatment	engender	hope
diagnose	problems	socialize	people
show	concern		

4.0 Hospitals/Health Care

4.1 Administration

determine	responsibility	diagnose	problems
schedule	operations	maintain	staff
comfort	patient	alert	specialists
educate	students	control	access
control	activities	prevent	contamination

5.0 Manufacturing

5.1 Equipment Assembly

ensure	safety	alter	components
create	product	identify	components
position	components	maintain	cleanliness
verify	operation	obtain	information
improve	appearance	prevent	damage

5.2 Environmental Improvements — Plants

fill	mold	blend	materials
release	heat	clean	casting
blend	materials	transport	material
seal	mold	release	gas
open	mold	melt	iron
contaminate	environment	create	heat

5.3 Heat Treating Operations

convert	energy	resist	wear
space	material	protect	personnel
minimize	distortion	control	noise
transport	energy	remove	waste

5.4 Foundry Operations

make	casting	reuse	flask
pour	iron	join	parts
complete	mold	control	quality
blend	materials	ensure	alignment
clean	product	store	materials

5.5 General Manufacturing Operation

load	components	obtain	approvals
rotate	assembly	estimate	manpower
deliver	material	control	cost
define	authority	determine	priority
clarify	operations	plan	effort
authorize	action	control	manpower

5.6 Production Bottleneck

improve	appearance	apply	finish
develop	uniformity	receive	material
transport	material	apply	information
identify	unit	protect	surface
assemble	components	control	noise

6.0 Personnel

6.1 Career Analysis & Planning

locate	client	maintain	standards
study	prospect	achieve	success
develop	plan	obtain	commitment
create	confidence	attract	attention
implement	plan	assure	progress

6.2 Job Evaluation and Design

arouse	interest	maintain	standards
show	benefit	measure	performance
obtain	commitment	utilize	capabilities
illustrate	potential	develop	relationship
evaluate	market	apply	skills

6.3 Organization Development — Drafting Productivity

define	product	highlight	deviations
obtain	information	involve	principles
develop	concept	increase	incentive
establish	guidelines	create	awareness
monitor	performance	reward	achievement

6.4   Position Description — Value Engineers

challenge	assumptions	make	decision
stimulate	creativity	ensure	results
integrate	skills	plan	approach
staff	operations	set	goals
analyze	capabilities	recommend	actions

7.0 Product Analysis

7.1   Air Conditioning and Ventilating Equipment

move	air	create	environment
convert	energy	control	noise
control	vibration	control	moisture
create	path	sense	requirement
direct	flow	achieve	comfort

7.2   Automobiles — Body and Components

create	environment	permit	entry
protect	operator	resist	bending
absorb	impact	seal	compartment
support	weight	locate	parts
transmit	load	conduct	air

7.3   Auto Parts and Assemblies

improve	appearance	guide	force
close	circuit	conduct	current
position	components	support	weight
transmit	torque	space	elements
reduce	drag	fail	safe
add	interest	attract	customer
generate	heat	collect	dirt
resist	corrosion	maintain	heater

7.4   Controls — Electrical/Mechanical

control	circuit	create	force
conduct	current	reduce	friction
apply	pressure	add	convenience
create	differential	rotate	component
extend	reach	multiply	force
direct	motion	convert	energy

7.5 Pumps and Compressors

create	flow	allow	access
increase	pressure	control	vibration
guide	fluid	connect	parts
create	noise	convert	energy
transfer	heat	seal	contents
contain	pressure	reduce	friction

7.6 Instrument Panel

transmit	information	store	material
distribute	impact	conduct	air
position	parts	regulate	climate
measure	performance	support	weight
create	impression	control	vehicle

7.7 Suspension Component

control	motion	support	weight
maintain	stability	transmit	forces
allow	adjustment	resist	corrosion
isolate	vibration	limit	travel

7.8 Telecommunication Equipment

accelerate	electrons	seal	system
position	components	connect	source
guide	signal	isolate	current
align	components	support	weight

7.9 Electrical Power Supply

create	heat	circulate	air
generate	noise	support	weight
position	components	resist	impact
protect	operator	resist	corrosion
isolate	current	allow	access

8.0 Product Planning/Marketing/Advertising

8.1 Future Vehicle Concept

8.2 High Impact, Low Cost Product Change

determine	needs	organize	service
develop	concept	evaluate	market
create	awareness	improve	balance
evaluate	concept	create	image
identify	market	stimulate	traffic
protect	investment	obtain	recognition

8.3 Long Lead Time Public Relations

attract	attention	illustrate	details
influence	public	complete	campaigns
generate	enthusiasm	prepare	schedule
demonstrate	product	educate	writers
collect	product	create	image

8.4 Auto Service in the Future

satisfy	customer	understand	requirements
develop	competence	create	confidence
fulfill	expectations	define	objectives
ensure	results	save	time
plan	program	improve	convenience

9.0 Real Estate/Building Management/Construction

9.1 Dealership Facility Concepts

exchange	goods	protect	assets
attract	prospect	store	merchandise
display	merchandise	identify	owner
manage	operation	train	personnel
protect	prospect	restore	agreement

9.2 Construction Program Integrated Into Corp. Plan

9.3 Laboratory Control and Operations Facility

verify	model	direct	operations
measure	performance	simulate	environment
obtain	parts	meet	criteria
prepare	model	collect	data
manage	personnel	evaluate	data

9.4 Construction Program/Scheduling

prepare	proposal	obtain	concurrence
analyze	conditions	resolve	restrictions
select	location	negotiate	agreement
obtain	data	authorize	expenditures
secure	option	monitor	progress
establish	budget	assign	responsibility

9.5 Construction Program Analysis

minimize	cost	disburse	funds
define	liabilities	ensure	compliance
protect	investment	educate	tenant
establish	standards	monitor	construction
evaluate	experience	prepare	forecast
contract	services	expedite	changes

10.0 Technical Operations

10.1 Development Dynamometers

maintain	schedule	establish	goals
educate	personnel	design	test
update	equipment	establish	priorities
transmit	Information	create	awareness
control	test	modify	designs
store	material	review	history
collect	data		

10.2 Emission Certification Facility

schedule	test	store	product
direct	personnel	move	product
maintain	equipment	measure	product
train	personnel	control	environment

10.3 Truck Engine Certification

construct	model	prepare	model
verify	performance	obtain	materiel
conserve	time	modify	equipment
obtain	tools	evaluate	line
collect	data	control	cost
store	data	compare	product

10.4 Vehicle Test and Development

confirm	design	resolve	problems
evaluate	performance	obtain	equipment
perform	service	identify	deviations
maintain	staff	utilize	facility
prepare	vehicle	control	cost
aid	development	store	information

10.5 Test Request Management

establish	priorities	evaluate	performance
evaluate	resources	generate	data
compare	information	evaluate	data
interpret	objectives	transmit	information
define	cost	develop	plans
review	history	set	schedule

10.6 Wind Tunnel Operations

obtain	data	measure	performance
monitor	equipment	verify	data
simulate	conditions	plan	operations
control	model	direct	personnel
prepare	model	control	operations
maintain	equipment	set	priorities

3. VERB–NOUN LIST
Function Verbs

A
absorb
accelerate
accept
achieve
acquire
actuate
add
adjust
administer
advise
aid
alert
align
allocate
allot
allow
alter
analyze
apply
approve
arouse
assemble
assess
assign
assist
assure
attract
authorize
award

B
blend
broaden
build

C
calibrate
certify

challenge
change
channel
charge
circulate
clarify
classify
clean
close
collect
comfort
communicate
compare
complete
compress
conduct
confirm
connect
conserve
consolidate
construct
consult
contain
contaminate
continue
contract
control
convert
convince
cool
coordinate
correct
cost
cover
create

D
decorate
decrease

define
delay
delegate
deliver
demonstrate
design
determine
develop
diagnose
direct
discharge
discuss
dispense
disperse
display
disseminate
dissipate
distribute
document

E
edit
educate
effect
eject
eliminate
employ
empty
enclose
encourage
engender
enhance
ensure
equalize
establish
estimate
evaluate
evangelize
exchange

excludes
exercise
exhaust
expand
expedite
explore
extend

F
fabricate
facilitate
fail
farm
fill
find
flow
forecast
form
foster
fulfill
furnish

G
gage
gather
generate
get
group
guide

H
highlight
hold

I
identify
illustrate
implement
improve

increase
indicate
indoctrinate
influence
inform
initiate
install
integrate
interpret
investigate
involve
isolate

J
join

L
latch
lease
limit
load
locate

M
maintain
make
manage
mandate
manufacture
market
match
maximize
measure
meet
melt
minimize
moderate
modify
monitor
motivate
mount
move

multiply

N
negotiate
nurture

O
observe
obtain
occupy
offer
open
operate
order
organize

P
package
partition
perform
permit
plan
position
pour
prepare
prescribe
present
prevent
prime
process
procure
produce
project
promote
protect

R
raise
receive
recognize
recommend
reconcile

redirect
reduce
reflect
regulate
release
remove
repair
replace
request
resist
resolve
respect
restore
retain
reuse
review
reward
rotate

S
satisfy
save
schedule
scrap
seal
secure
seek
select
sell
sense
separate
serve
service
set
settle
share
show
signal
socialize
solicit
solve
space

specify
speed
spray
staff
stimulate
stimulate
store
stow
study
summarize
supply
support

T
teach
terminate
test
track
train
transcribe
transfer
transmit
transport

U
understand
update
use
utilize

V
validate
vary
verify
visit

W
welcome
worship

Function Nouns

A
abilities
acceptability
access
accountability
accuracy
achievement
action
activities
adjustment
administration
affection
agencies
agreement
air
alignment
alternatives
appearance
approach
area
arrival
assembly
assets
associates
associations
assumptions
attention
attire
attitudes
authority
awareness

B
balance
behavior
beliefs
bending
benefits
bids
binder
breakthrough

budget
buyer

C
campaign
capabilities
casting
catalog
change
checklist
children
circuit
claim
classes
clearance
client
climate
comfort
commitment
committees
community
companionship
comparison
compartment
competence
compliance
component
concept
concern
concurrence
condition
confession
confidence
conflict
conformance
construction
contact
contamination
contents
contingencies
contractor

contracts
contrast
convenience
cope
core
corrosion
cost
council
creativity
criteria
current
customer

D
damage
data
dealer
decision
deflection
demand
design
details
development
deviancy
deviation
differences
differential
dimension
direction
dirt
distortions
distribution
document
documentation
downtime
drag
duplication

E
efficiency
effort

electrons
elements
emissions
emotions
empathy
energy
enthusiasm
entry
environment
equipment
errors
event
expectations
expenditures
experience

F
facility
failure
feasibility
feedback
fellowship
financing
flask
flexibility
float
flow
fluid
force
forecast
freedom
friction
friendship
fumes
funds

G
gas
goals
God
good

goods
growth
guidance
guidelines

H
habits
haven
healing
health
heat
heater
history
hope

I
idea
identification
illumination
image
impact
impression
impurities
incentive
income
information
injury
interest
interface
inventory
investment
iron

J
justice

L
labor
lagoon
land
launch
law

layout
level
liability
liaison
life
light
limit
line
liturgy
load
location
luster

M
management
manpower
market
material
maturation
measurement
merchandise
method
mixture
model
moisture
mold
motion
motor
movement

N
need
noise
notion

O
objectives
obligation
offer
opening
operation
operator

opinion
opportunity
option
order
organization
others
owners

P
pace
package
participation
parts
passage
passenger
path
patient
pattern
payment
people
performance
personnel
piece
plan
policies
pool
potential
power
prerogative
pressure
price
principles
priorities
privacy
problem
procedure
process
product
production
profit
program
progress

projection
property
proposal
prospect
public

Q
Quality

R
rating
reach
recipe
recognition
recommendation
records
relations
relationship
repair
report
request
requirements
resources
respect
responsibility
restrictions
results
rigidity
risk
rotation
rules
runout

S
safe
safety
sales
sand
satisfaction
schedule
scriptures
series

service
shaker
shock
signal
skill
solace
solids
sounds
sources
space
specialists
specifications
stability
staff
standards
stewardship
strangers
stress
structure

students
study
success
suggestions
suppliers
supply
surface
surplus
systems

T
team
teamwork
temperature
tenant
terms
test
thought
time

timing
torque
tour
tradition
traffic
travel
treatment
trust

U
understanding
unique
user
utilities

V
vacuum
value
vapor

vehicle
vendor
vibration
views
visibility
volume
volunteers

W
warranty
waste
water
wear
weight
work
workload
writers

Worksheets

Information Phase   Sheet 1

Project
Information   _____ Date

Part Name	Drawing or part no

Used on (Name or Number)	Number required per assembly

Team number	Workshop number and date	Task force and date

Team Members	Department	Phone number

Present cost

Total cost	Cost Elements		
	Material	Labor	Burden

Estimated annual production

Operation and Performance

Includes other costs - See sheet 3

1. Information Phase   Sheet 2

Determine Scope
Identify Functions _____ Date

Project Name	Drawing/Part No.
What is it?	Scope includes
Other limits of project.	Scope does not include
What does it do?	

List all functions without constraints

	Verb	Noun	Basic	Second	Remarks
1.					
2.					
3.					
4.					
5.					
6.					
7.					
8.					
9.					
10.					
11.					
12.					
13.					
14.					
15.					
16.					
17.					
18.					
19.					
20.					
21.					
22.					
23.					
24.					
25.					
26.					
27.					
28.					
29.					
30.					

Determine Scope
Identify Functions _____ Date

Project Name	Drawing/Part No.
What is it?	Scope includes
Other limits of project.	Scope does not include
What does it do?	

List all functions without constraints

	Verb	Noun	Basic	Second	Remarks
1.					
2.					
3.					
4.					
5.					
6.					
7.					
8.					
9.					
10.					
11.					
12.					
13.					
14.					
15.					
16.					
17.					
18.					
19.					
20.					
21.					
22.					
23.					
24.					
25.					
26.					
27.					
28.					
29.					
30.					

II.   **Creative Phase**   Sheet 4

Develop Ideas

_____Date

1.
2.
3.
4.
5.
6.
7.
8.
9.
10.
11.
12.
13.
14.
15.
16.
17.
18.
19.
20.
21.
22.
23.
24.
25.
26.
27.
28.
29.
30.
31.
32.
33.
34.
35.
36.
37.
1.

Screen All

Ideas _____ Date

	Idea	Cost per	Tool cost	Develop. cost	Total cost
1.					
2.					
3.					
4.					
5.					
6.					
7.					
8.					
9.					
10.					
11.					
12.					
13.					
14.					
15.					
16.					
17.					
18.					
19.					
20.					
21.					
22.					
23.					
24.					
25.					
26.					
27.					
28.					
29.					
30.					

3. **Evaluation Phase**   sheet 6

Select
Best Ideas _____ Date

	Idea		Idea	
	Benefits	Disadvantages	Benefits	Disadvantages
1.				
2.				
3.				
4.				
5.				
6.				
7.				
8.				
9.				
10				
11				
12				
13				
14				
15				

	Idea		Idea	
	Benefits	Disadvantages	Benefits	Disadvantages
1.				
2.				
3.				
4.				
5.				
6.				
7.				
8.				
9.				
10				
11				
12				
13				
14				
15				

Function Value - Best alternative $\$$_____   Measure of Value $= \dfrac{function\ value}{present\ cost} \times 100 =$ ____%

Present Cost $\$$_____ −

4. Planning Phase   **Sheet 7**

Identify
Roadblocks   _____Date

Select the best ideas

Roadblock	Where / Why ?	Action required

Alternative idea

Roadblock	Where / Why	Action required

4. Planning Phase   Sheet 8

How Do I
Sell It ?   _____Date

Implementation document - Determine all data required. Attach additional sheets if necessary.
Persons and areas involved - List all departments and organizations that will be required to do something as a result of this proposal and the names and supervisors of those directly involved.

	Depart.	Supervisor	Action	Cost	
				Material	Labor
1.					
2.					
3.					
4.					
5.					
6.					
7.					
8.					
9.					
10					
11					
12					

Potential problem areas - Indicate any foreseeable problem area or unusual situation that might result during the implementation of the proposal

	Dept.	Problem
1.		
2.		
3.		
4.		
5.		
6.		
7.		
8.		
9.		
10		
11		
12		

4. Planning Phase   Sheet 9

Project Study
Record   _____ Date

Minimize Problem areas — How might the problem areas identified above be minimized ?

	Dept.	Problem Area	How to minimize
1.			
2.			
3.			
4.			
5.			
6.			
7.			
8.			
9.			
10.			
11.			
12.			
13.			
14.			
15.			

General verbal approval obtained — Preliminary discussions with people directly involved will usually pave the way for quick formal approval of the implementation document. Note the result of the discussions here.

Name	Comments	Approval	Date
Principle decision maker			
Others directly involved			

4. Planning Phase   Sheet 10

**Notes and
Sketches**   _____date

Be sure to date and witness important ideas for future patent action.

1 inch X 1 inch squares

Value Control   Sheet 11
V. Reporting Phase

Change   Date
Proposal   _____

Example   1

Product
_____

Part Assembly
_____

Reference
_____

Description of Charge

Present	Proposed

Advantages

Disadvantages

Cost Analysis									
Implimention			Variable Cost per piece					Potential Saving	
	Material	Labor		Material	labor	Burden	Total	Benefits are based	
Eng.			Present					on prompt approval	
Dvp/			Prop.					Series	Net Savin
Test			1. Pc. Cost saving						
Tooling			2. Est. Number units						
Equip.			3. Gross saving (1 x 2)						
			4. Total Implementation Costs						
			5. Net saving 1st year (3 - 4)						
Total			6. Payout - Months 4/3 x 12					Life	
Total Material & labor									

Potential Profit Improvement Effect

POTENTIAL PROFIT IMPROVEMENT EFFECT

PIECE COST SAVING　　　　　$\boxed{}$

NUMBER OF UNITS/YR.　　　　　_____

GROSS SAVING　　　　　　　_____
NET SAVING 1st YEAR　　　$\boxed{}$

NET SAVING LIFETIME　　　　$\boxed{}$

PAYOUT ___ **MONTHS, TARGET DATE** _____

Index